THE OBESITY CODE

Dr Jason Fung grew up in Toronto, Canada and completed both medical school and an internal medicine residency at the University of Toronto. He headed to the University of California, Los Angeles completing his fellowship in nephrology (kidney disease specialist). He now has both a hospital- and office-based practice in Toronto, and is the current chief of the Department of Medicine at The Scarborough Hospital, General Division.

Struggling daily against the worsening epidemic of type 2 diabetes and obesity, Dr Fung realised that current recommended treatment of 'Eat Less, Move More' was simply not successful. It soon became clear that the medical obsession with calories was not the proper model to treat obesity. He established the Intensive Dietary Management Program to provide patients with a unique treatment focus on hormones rather than diet. The program treats conditions related to metabolic syndrome, including obesity, type 2 diabetes, obstructive sleep apnea, and fatty liver with great success. It now provides guidance both locally and to international patients from as far as New Zealand to the United Kingdom to South Africa.

Dr Fung lives in Toronto with his wife and two boys.

THE OBESITY CODE

UNLOCKING THE SECRETS OF WEIGHT LOSS

JASON FUNG

Melbourne • London

Scribe Publications
2 John St, Clerkenwell, London, WC1N 2ES, United Kingdom
18–20 Edward St, Brunswick, Victoria 3056, Australia

Published by Scribe 2016
Reprinted 2016 (twice), 2017, 2018, 2019 (twice), 2020

Figure 4.1 in page 50 is used with permission of Public Health England. Figure 12.1 on page 137 is used with the permission of the CDC. The use of this figure does not constitute endorsement by the CDC. Figure 14.1 on page 163 is used with permission of Dr George Bray. All other figures are copyright Jason Fung. Some of this material appeared previously on the Intensive Dietary Management website: www.intensivedietarymanagement.com.

Editing by Eva van Emden
Cover and text design by Nayeli Jimenez

Printed and bound in the UK by CPI Group (UK) Ltd, Croydon CR0 4YY

Scribe Publications is committed to the sustainable use of natural resources and the use of paper products made responsibly from those resources.

9781925228793 (UK paperback)
9781925321517 (Australian paperback)
9781925307689 (e-book)

Catalogue records for this book are available from the British Library and the National Library of Australia.

scribepublications.co.uk
scribepublications.com.au

This book is dedicated to my beautiful wife, Mina.
Thank you for your love and the strength you give me. I could not do it without you, nor would I ever want to.

CONTENTS

FOREWORD

•

DR. JASON FUNG is a Toronto physician specializing in the care of patients with kidney diseases. His key responsibility is to oversee the complex management of patients with end-stage kidney disease requiring renal (kidney) dialysis.

His credentials do not obviously explain why he should author a book titled *The Obesity Code* or why he blogs on the intensive dietary management of obesity and type 2 diabetes mellitus. To understand this apparent anomaly, we need first to appreciate who this man is and what makes him so unusual.

In treating patients with end-stage kidney disease, Dr. Fung learned two key lessons. First, that type 2 diabetes is the single commonest cause of kidney failure. Second, that renal dialysis, however sophisticated and even life prolonging, treats only the final symptoms of an underlying disease that has been present for twenty, thirty, forty or perhaps even fifty years. Gradually, it dawned on Dr. Fung that he was practicing medicine exactly as he had been taught: by reactively treating the symptoms of complex diseases without first trying to understand or correct their root causes.

He realized that to make a difference to his patients, he would have to start by acknowledging a bitter truth: that our venerated profession is no longer interested in addressing the causes of disease. Instead, it wastes much of its time and many of its resources attempting to treat symptoms.

He resolved to make a real difference to his patients (and his profession) by striving to understand the true causes that underlie disease.

Before December 2014, I was unaware of Dr. Jason Fung's existence. Then one day I chanced upon his two lectures—'The Two Big Lies of Type 2 Diabetes' and 'How to Reverse Type 2 Diabetes Naturally'—on YouTube. As someone with a special interest in type 2 diabetes, not least because I have the condition myself, I was naturally intrigued. Who, I thought, is this bright young man? What gives him the certainty that type 2 diabetes can be reversed 'naturally'? And how can he be brave enough to accuse his noble profession of lying? He will need to present a good argument, I thought.

It took only a few minutes to realize that Dr. Fung is not only legitimate, but also more than able to look after himself in any medical scrap. The argument he presented was one that had been bouncing around, unresolved, in my own mind for at least three years. But I had never been able to see it with the same clarity or to explain it with the same emphatic simplicity as had Dr. Fung. By the end of his two lectures, I knew that I had observed a young master at work. Finally, I understood what I had missed.

What Dr. Fung achieved in those two lectures was to utterly destroy the currently popular model for the medical management of type 2 diabetes—the model mandated by all the different diabetes associations around the world. Worse, he explained why this erroneous model of treatment must inevitably harm the health of all patients unfortunate enough to receive it.

According to Dr. Fung, the first big lie in the management of type 2 diabetes is the claim that it is a chronically progressive disease that simply gets worse with time, even in those who comply with the best treatments modern medicine offers. But, Dr. Fung argues, this is simply not true. Fifty per cent of the patients on Dr. Fung's Intensive Dietary Management (IDM) program, which combines dietary carbohydrate restriction and fasting, are able to stop using insulin after a few months.

So why are we unable to acknowledge the truth? Dr. Fung's answer is simple: we doctors lie to ourselves. If type 2 diabetes is a curable disease but all our patients are getting worse on the treatments we prescribe, then we must be bad doctors. And since we did not study for so long at such great cost to become bad doctors, this failure cannot be our fault. Instead, we must believe we are doing the best for our patients, who must unfortunately be suffering from a chronically progressive and incurable disease. It is not a deliberate lie, Dr. Fung concludes, but one of cognitive dissonance—the inability to accept a blatant truth because accepting it would be too emotionally devastating.

The second lie, according to Dr. Fung, is our belief that type 2 diabetes is a disease of abnormal blood glucose levels for which the only correct treatment is progressively increasing insulin dosages. He argues, instead, that type 2 diabetes is a disease of insulin resistance with *excessive* insulin secretion—in contrast to type 1 diabetes, a condition of true insulin *lack*. To treat both conditions the same way—by injecting insulin—makes no sense. Why treat a condition of insulin excess with yet more insulin, he asks? That is the equivalent of prescribing alcohol for the treatment of alcoholism.

Dr. Fung's novel contribution is his insight that treatment in type 2 diabetes focuses on the symptom of the disease—an elevated blood glucose concentration—rather than its root cause, insulin resistance. And the initial treatment for insulin resistance is to limit carbohydrate intake. Understanding this simple biology explains why this disease may be reversible in some cases—and, conversely, why the modern treatment of type 2 diabetes, which does not limit carbohydrate intake, worsens the outcome.

But how did Dr. Fung arrive at these outrageous conclusions? And how did they lead to his authorship of this book?

In addition to his realization, described above, of the long-term nature of disease and the illogic of treating a disease's symptoms rather than removing its cause, he also, almost by chance, in the early 2000s, became aware of the growing literature on the benefits of low-carbohydrate diets in those with obesity and other conditions of insulin resistance. Taught to believe that a carbohydrate-restricted, high-fat diet kills, he was shocked to discover the opposite: this dietary choice produces a range of highly beneficial metabolic outcomes, especially in those with the worst insulin resistance.

And finally came the cherry on the top—a legion of hidden studies showing that for the reduction of body weight in those with obesity (and insulin resistance), this high-fat diet is at least as effective, and usually much more so, than other more conventional diets.

Eventually, he could bear it no longer. If everyone knows (but won't admit) that the low-fat calorie-restricted diet is utterly ineffective in controlling body weight or in treating obesity, surely it is time to tell the truth: the best hope for treating and preventing obesity, a disease of insulin resistance and excessive insulin production, must surely be the same low-carbohydrate, high-fat diet used for the management of the ultimate disease of insulin resistance, type 2 diabetes. And so this book was born.

In *The Obesity Code,* Dr. Fung has produced perhaps the most important popular book yet published on this topic of obesity.

Its strengths are that it is based on an irrefutable biology, the evidence for which is carefully presented; and it is written with the ease and confidence of a master communicator in an accessible, well-reasoned sequence so that its consecutive chapters systematically develop, layer by layer, an evidence-based biological model of obesity that makes complete sense in its logical simplicity. It includes just enough science to convince the skeptical scientist, but not so much that it confuses those without a background in biology. This feat in itself is a stunning achievement that few science writers ever accomplish.

By the end of the book, the careful reader will understand exactly the causes of the obesity epidemic, why our attempts to prevent both the obesity and diabetes epidemics were bound to fail, and what, more importantly, are the simple steps that those with a weight problem need to take to reverse their obesity.

The solution needed is that which Dr. Fung has now provided: 'Obesity is... a multifactorial disease. What we need is a framework, a structure, a coherent theory to understand how all its factors fit together. Too often, our current model of obesity assumes that there is only one single true cause, and that all others are pretenders to the throne. Endless debates ensue... They are all partially correct.'

In providing one such coherent framework that can account for most of what we currently know about the real causes of obesity, Dr. Fung has provided much, much more.

He has provided a blueprint for the reversal of the greatest medical epidemics facing modern society—epidemics that he shows are entirely preventable and potentially reversible, but only if we truly understand their biological causes—not just their symptoms.

The truth he expresses will one day be acknowledged as self-evident. The sooner that day dawns, the better for us all.

TIMOTHY NOAKES OMS, MBChB, MD, DSc, PhD (hc), FACSM, (hon) FFSEM (UK), (hon) FSEM (Ire)
Emeritus Professor
University of Cape Town, Cape Town, South Africa

INTRODUCTION

•

THE ART OF medicine is quite peculiar. Once in a while, medical treatments become established that don't really work. Through sheer inertia, these treatments get handed down from one generation of doctors to the next and survive for a surprisingly long time, despite their lack of effectiveness. Consider the medicinal use of leeches (bleeding) or, say, routine tonsillectomy.

Unfortunately, the treatment of obesity is also one such example. Obesity is defined in terms of a person's body mass index, calculated as a person's weight in kilograms divided by the square of their height in meters. A body mass index greater than 30 is defined as obese. For more than thirty years, doctors have recommended a low-fat, calorie-reduced diet as the treatment of choice for obesity. Yet the obesity epidemic accelerates. From 1985 to 2011, the prevalence of obesity in Canada tripled, from 6 per cent to 18 per cent.[1] This phenomenon is not unique to North America, but involves most of the nations of the world.

Virtually every person who has used caloric reduction for weight loss has failed. And, really, who hasn't tried it? By every objective measure, this treatment is completely and utterly ineffective. Yet it remains the treatment of choice, defended vigorously by nutritional authorities.

As a nephrologist, I specialize in kidney disease, the most common cause of which is type 2 diabetes with its associated obesity. I've often watched patients start insulin treatment for their diabetes, knowing that most will gain weight. Patients are rightly concerned. 'Doctor,' they say, 'you've always told me to lose weight. But the insulin you gave me makes me gain so much weight. How is this helpful?' For a long time, I didn't have a good answer for them.

That nagging unease grew. Like many doctors, I believed that weight gain was a caloric imbalance—eating too much and moving too little. But if that were so, why did the medication I prescribed—insulin—cause such relentless weight gain?

Everybody, health professionals and patients alike, understood that the root cause of type 2 diabetes lay in weight gain. There were rare cases of highly motivated patients who had lost significant amounts of weight. Their type 2 diabetes would also reverse course. Logically, since weight was the underlying problem, it deserved significant attention. Still, it seemed that the health profession was not even the least bit interested in treating it. I was guilty as charged. Despite having worked for more than twenty years in medicine, I found that my own nutritional knowledge was rudimentary, at best.

Treatment of this terrible disease—obesity—was left to large corporations like Weight Watchers, as well as various hucksters and charlatans mostly interested in peddling the latest weight-loss 'miracle.' Doctors were not even remotely interested in nutrition. Instead, the medical profession seemed obsessed with finding and prescribing the next new drug:

- You have type 2 diabetes? Here, let me give you a pill.
- You have high blood pressure? Here, let me give you a pill.
- You have high cholesterol? Here, let me give you a pill.
- You have kidney disease? Here, let me give you a pill.

But all along, *we needed to treat obesity.* We were trying to treat the problems caused by obesity rather than obesity itself. In trying to understand the underlying cause of obesity, I eventually established the Intensive Dietary Management Clinic in Toronto, Canada.

The conventional view of obesity as a caloric imbalance did not make sense. Caloric reduction had been prescribed for the last fifty years with startling ineffectiveness.

Reading books on nutrition was no help. That was mostly a game of 'he said, she said,' with many quoting 'authoritative' doctors. For example, Dr. Dean Ornish says that dietary fat is bad and carbohydrates are good. He is a respected doctor, so we should listen to him. But Dr. Robert Atkins said dietary fat is good and carbohydrates are bad. He was also a respected doctor, so we should listen to him. Who is right? Who is wrong? In the science of nutrition, there is rarely any consensus about *anything*:

- Dietary fat is bad. No, dietary fat is good. There are good fats and bad fats.
- Carbohydrates are bad. No, carbohydrates are good. There are good carbs and bad carbs.
- You should eat more meals a day. No, you should eat fewer meals a day.
- Count your calories. No, calories don't count.
- Milk is good for you. No, milk is bad for you.
- Meat is good for you. No, meat is bad for you.

To discover the answers, we need to turn to evidence-based medicine rather than vague opinion.

Literally thousands of books are devoted to dieting and weight loss, usually written by doctors, nutritionists, personal trainers and other 'health experts.' However, with a few exceptions, rarely is more than a cursory thought spared for the actual *causes* of obesity. What *makes* us gain weight? Why do we get fat?

The major problem is the complete lack of a theoretical framework for understanding obesity. Current theories are ridiculously simplistic, often taking only one factor into account:

- Excess calories cause obesity.
- Excess carbohydrates cause obesity.
- Excess meat consumption causes obesity.
- Excess dietary fat causes obesity.
- Too little exercise causes obesity.

But all chronic diseases are multifactorial, and these factors are not mutually exclusive. They may all contribute to varying degrees. For example, heart disease has numerous contributing factors—family history, gender, smoking, diabetes, high cholesterol, high blood pressure and a lack of physical activity, to name only a few—and that fact is well accepted. But such is not the case in obesity research.

The other major barrier to understanding is the focus on short-term studies. Obesity usually takes decades to fully develop. Yet we often rely on information about it from studies that are only of several weeks' duration. If we study how rust develops, we would need to observe metal over a period of weeks to months, not hours. Obesity, similarly, is a long-term disease. Short-term studies may not be informative.

While I understand that the research is not always conclusive, I hope this book, which draws on what I've learned over twenty years of helping patients with type 2 diabetes lose weight permanently to manage their disease, will provide a structure to build upon.

Evidence-based medicine does not mean taking every piece of low-quality evidence at face value. I often read statements such as 'low-fat diets proven to completely reverse heart disease.' The reference will be a study of five rats. That hardly qualifies as evidence. I will reference only studies done on humans, and mostly only those that have been published in high-quality, peer-reviewed journals. No animal studies will be discussed in this book. The reason for this decision can be illustrated in 'The Parable of the Cow':

Two cows were discussing the latest nutritional research, which had been done on lions. One cow says to the other, 'Did you hear that we've been wrong these last 200 years? The latest research shows that eating grass is bad for you and eating meat is good.' So the two cows began eating meat. Shortly afterward, they got sick and they died.

One year later, two lions were discussing the latest nutritional research, which was done on cows. One lion said to the other that the latest research showed that eating meat kills you and eating grass is good. So, the two lions started eating grass, and they died.

What's the moral of the story? We are not mice. We are not rats. We are not chimpanzees or spider monkeys. We are human beings, and therefore we should consider only human studies. I am interested in obesity in humans, not obesity in mice. As much as possible, I try to focus on causal factors rather than association studies. It is dangerous to assume that because two factors are associated, one is the cause of the other. Witness the hormone replacement therapy disaster in post-menopausal women. Hormone replacement therapy was *associated* with lower heart disease, but that did not mean that it was the *cause* of lower heart disease. However, in nutritional research, it is not always possible to avoid association studies, as they are often the best available evidence.

Part 1 of this book, 'The Epidemic,' explores the timeline of the obesity epidemic and the contribution of the patient's family history, and shows how both shed light on the underlying causes.

Part 2, 'The Calorie Deception,' reviews the current caloric theory in depth, including exercise and overfeeding studies. The shortcomings of the current understanding of obesity are highlighted.

Part 3, 'A New Model of Obesity,' introduces the hormonal theory of obesity, a robust explanation of obesity as a medical problem. These chapters explain the central role of insulin in regulating body weight and describe the vitally important role of insulin resistance.

Part 4, 'The Social Phenomenon of Obesity,' considers how hormonal obesity theory explains some of the associations of obesity. Why is obesity associated with poverty? What can we do about childhood obesity?

Part 5, 'What's Wrong with Our Diet?,' explores the role of fat, protein and carbohydrates, the three macronutrients, in weight gain. In addition, we examine one of the main culprits in weight gain—fructose—and the effects of artificial sweeteners.

Part 6, 'The Solution,' provides guidelines for lasting treatment of obesity by addressing the hormonal imbalance of high blood insulin. Dietary guidelines for reducing insulin levels include reducing added sugar and refined grains, keeping protein consumption moderate, and adding healthy fat and fiber. Intermittent fasting is an effective way to treat insulin resistance without incurring the negative effects of calorie reduction diets. Stress management and sleep improvement can reduce cortisol levels and control insulin.

The Obesity Code will set forth a framework for understanding the condition of human obesity. While obesity shares many important similarities and differences with type 2 diabetes, this is primarily a book about obesity.

The process of challenging current nutritional dogma is, at times, unsettling, but the health consequences are too important to ignore. What actually causes weight gain and what can we do about it? This question is the overall theme of this book. A fresh framework for the understanding and treatment of obesity represents a new hope for a healthier future.

JASON FUNG, MD

PART ONE

The Epidemic

(1)

HOW OBESITY BECAME AN EPIDEMIC

•

Of all the parasites that affect humanity, I do not know of, nor can I imagine, any more distressing than that of Obesity.
WILLIAM BANTING

HERE'S THE QUESTION that has always bothered me: Why are there doctors who are fat? Accepted as authorities in human physiology, doctors should be true experts on the causes and treatments of obesity. Most doctors are also very hardworking and self-disciplined. Since nobody wants to be fat, doctors in particular should have both the knowledge and the dedication to stay thin and healthy.

So why are there fat doctors?

The standard prescription for weight loss is 'Eat Less, Move More.' It *sounds* perfectly reasonable. But why doesn't it work? Perhaps people wanting to lose weight are not following this advice. The mind is willing, but the flesh is weak. Yet consider the self-discipline and dedication needed to complete an undergraduate degree, medical school, internship, residency and fellowship. It is hardly conceivable that overweight doctors simply lack the willpower to follow their own advice.

This leaves the possibility that the conventional advice is simply wrong. And if it is, then our entire understanding of obesity is fundamentally flawed. Given the current epidemic of obesity, I suspect that such is the most likely scenario. So we need to start at the very beginning, with a thorough understanding of the disease that is human obesity.

We must start with the single most important question regarding obesity or any disease: 'What causes it?' We spend no time considering this crucial question because we think we already know the answer. It seems so obvious: it's a matter of Calories In versus Calories Out.

A calorie is a unit of food energy used by the body for various functions such as breathing, building new muscle and bone, pumping blood and other metabolic tasks. Some food energy is stored as fat. Calories In is the food energy that we eat. Calories Out is the energy expended for all of these various metabolic functions.

When the number of calories we take in exceeds the number of calories we burn, weight gain results, we say. Eating too much and exercising too little causes weight gain, we say. Eating too many *calories* causes weight gain, we say. These 'truths' seem so self-evident that we do not question whether they are actually true. But are they?

PROXIMATE VERSUS ULTIMATE CAUSE

EXCESS CALORIES MAY certainly be the *proximate* cause of weight gain, but not its *ultimate* cause.

What's the difference between proximate and ultimate? The proximate cause is *immediately* responsible, whereas the ultimate cause is what started the chain of events.

Consider alcoholism. What causes alcoholism? The proximate cause is 'drinking too much alcohol'—which is undeniably true, but not particularly useful. The question and the cause here are one and the same, since alcoholism *means* 'drinking too much alcohol.' Treatment advice directed against the proximate cause—'Stop drinking so much alcohol'—is not useful.

The crucial question, the one that we are really interested in, is: What is the *ultimate* cause of *why* alcoholism occurs. The ultimate cause includes

- the addictive nature of alcohol,
- any family history of alcoholism,
- excessive stress in the home situation and/or
- an addictive personality.

There we have the real disease, and treatment must be directed against the ultimate, rather than the proximate cause. Understanding the ultimate cause leads to effective treatments such as (in this case) rehabilitation and social support networks.

Let's take another example. Why does a plane crash? The proximate cause is, 'there was not enough lift to overcome gravity'—again, absolutely true, but not in any way useful. The ultimate cause might be

- human error,
- mechanical fault and/or
- inclement weather.

Understanding the ultimate cause leads to effective solutions such as better pilot training or tighter maintenance schedules. Advice to 'generate more lift than gravity' (larger wings, more powerful engines) will not reduce plane crashes.

This understanding applies to everything. For instance, why is it so hot in this room?

PROXIMATE CAUSE: Heat energy coming in is greater than heat energy leaving.

SOLUTION: Turn on the fans to increase the amount of heat leaving.

ULTIMATE CAUSE: The thermostat is set too high.

SOLUTION: Turn down the thermostat.

Why is the boat sinking?

PROXIMATE CAUSE: Gravity is stronger than buoyancy.

SOLUTION: Reduce gravity by lightening the boat.

ULTIMATE CAUSE: The boat has a large hole in the hull.

SOLUTION: Patch the hole.

In each case, the solution to the proximate cause of the problem is neither lasting nor meaningful. By contrast, treatment of the ultimate cause is far more successful.

The same applies to obesity: What causes weight gain?

Proximate cause: Consuming more calories than you expend.

If more calories in than out is the proximate cause, the unspoken answer to that last question is that the ultimate cause is 'personal choice.' We *choose* to eat chips instead of broccoli. We *choose* to watch TV instead of exercise. Through this reasoning, obesity is transformed from a disease that needs to be investigated and understood into a personal failing, a character defect. Instead of searching for the ultimate cause of obesity, we transform the problem into

- eating too much (gluttony) and/or
- exercising too little (sloth).

Gluttony and sloth are two of the seven deadly sins. So we say of the obese that they 'brought it on themselves.' They 'let themselves go.' It gives us the comforting illusion that we understand ultimate cause of the problem. In a 2012 online poll, 61 per cent of U.S. adults believed that 'personal choices about eating and exercise' were responsible for the obesity epidemic.[1] So we discriminate against people who are obese. We both pity and loathe them.

However, on simple reflection, this idea simply cannot be true. Prior to puberty, boys and girls average the same body-fat percentage. After puberty, women on average carry close to 50 per cent more body fat than men. This change occurs despite the fact that men consume more calories on average than women. But why is this true?

What is the ultimate cause? It has nothing to do with personal choices. It is not a character defect. Women are not more gluttonous or lazier than men. The hormonal cocktail that differentiates men and women must make it more likely that women will accumulate excess calories as fat as opposed to burning them off.

Pregnancy also induces significant weight gain. What is the ultimate cause? Again, it is obviously the hormonal changes resulting from the pregnancy—*not* personal choice—that encourages weight gain.

Having erred in understanding the proximate and ultimate causes, we believe the solution to obesity is to eat fewer calories.

The 'authorities' all agree. The U.S. Department of Agriculture's *Dietary Guidelines for Americans,* updated in 2010, forcefully proclaims its key recommendation: 'Control total calorie intake to manage body weight.' The Centers for Disease Control exhort patients to balance their calories.[2] The advice from the National Institutes of Health's pamphlet 'Aim for a Healthy Weight' is 'to cut down on the number of calories . . . they get from food and beverages and increase their physical activity.'[3]

All this advice forms the famous 'Eat Less, Move More' strategy so beloved by obesity 'experts.' But here's a peculiar thought: If we already understand what causes obesity, how to treat it, and we've spent millions of dollars on education and obesity programs, *why are we getting fatter?*

ANATOMY OF AN EPIDEMIC

WE WEREN'T ALWAYS so obsessed with calories. Throughout most of human history, obesity has been rare. Individuals in traditional societies eating traditional diets seldom became obese, even in times of abundant food. As civilizations developed, obesity followed. Speculating on the cause, many identified the refined carbohydrates of sugar and starches. Sometimes considered the father of the low-carbohydrate diet, Jean Anthelme Brillat-Savarin (1755–1826) wrote the influential textbook *The Physiology of Taste* in 1825. There he wrote: 'The second of the chief causes of obesity is the *floury and starchy substances* which man makes the prime ingredients of his daily nourishment. As we have said already, all animals that live on farinaceous food grow fat willy-nilly; and man is no exception to the universal law.'[4]

All foods can be divided into three different macronutrient groups: fat, protein and carbohydrates. The 'macro' in 'macronutrients' refers to the fact that the bulk of the food we eat is made up of these three groups. Micronutrients, which make up a very small proportion of the food, include vitamins and minerals such as vitamins A, B, C, D, E and K, as well as minerals such as iron and calcium. Starchy foods and sugars are all carbohydrates.

Several decades later, William Banting (1796–1878), an English undertaker, rediscovered the fattening properties of the refined carbohydrate. In 1863, he published the pamphlet *Letter on Corpulence, Addressed to the Public,* which is often considered the world's first diet book. His story is rather unremarkable. He was not an obese child, nor did he have a family history of obesity. In his mid-thirties, however, he started to gain weight. Not much; perhaps a pound or two per year. By age sixty-two, he stood five foot five and weighed 202 pounds (92 kilograms). Perhaps unremarkable by modern standards, he was considered quite portly at the time. Distressed, he sought advice on weight loss from his physicians.

First, he tried to eat less, but that only left him hungry. Worse, he failed to lose weight. Next, he increased his exercise by rowing along the River Thames, near his home in London. While his physical fitness improved, he developed a 'prodigious appetite, which I was compelled to indulge.'[5] Still, he failed to lose weight.

Finally, on the advice of his surgeon, Banting tried a new approach. With the idea that sugary and starchy foods were fattening, he strenuously avoided all breads, milk, beer, sweets and potatoes that had previously made up a large portion of his diet. (Today we would call this diet low in refined carbohydrates.) William Banting not only lost the weight and kept it off, but he also felt so well that he was compelled to write his famous pamphlet. Weight gain, he believed, resulted from eating too many 'fattening carbohydrates.'

For most of the next century, diets low in refined carbohydrates were accepted as the standard treatment for obesity. By the 1950s, it was fairly standard advice. If you were to ask your grandparents what caused obesity, they would not talk about calories. Instead, they would tell you to stop eating sugary and starchy foods. Common sense and empiric observation served to confirm the truth. Nutritional 'experts' and government opinion were not needed.

Calorie counting had begun in the early 1900s with the book *Eat Your Way to Health,* written by Dr. Robert Hugh Rose as a 'scientific system of weight control.' That book was followed up in 1918 with the bestseller *Diet and Health, with Key to the Calories,* written by Dr. Lulu Hunt Peters, an American doctor and newspaper columnist. Herbert Hoover, then the head of the U.S. Food Administration, converted to calorie counting. Dr. Peters advised patients to start with a fast, one to two days abstaining from all foods, and then stick strictly to 1200 calories per day. While the advice to fast was quickly forgotten, modern calorie-counting schedules are not very different.

By the 1950s, a perceived 'great epidemic' of heart disease was becoming an increasing public concern. Seemingly healthy Americans were developing heart attacks with growing regularity. In hindsight, it should have been obvious that there was really no such epidemic.

The discovery of vaccines and antibiotics, combined with increased public sanitation, had reshaped the medical landscape. Formerly lethal infections, such as pneumonia, tuberculosis and gastrointestinal infections, became curable. Heart disease and cancer now caused a relatively greater percentage of deaths, giving rise to some of the public misperception of an epidemic. (See Figure 1.1.[6])

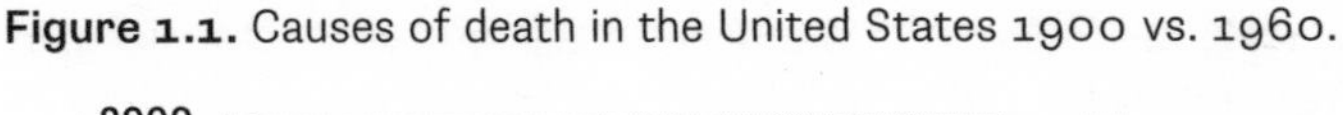

Figure 1.1. Causes of death in the United States 1900 vs. 1960.

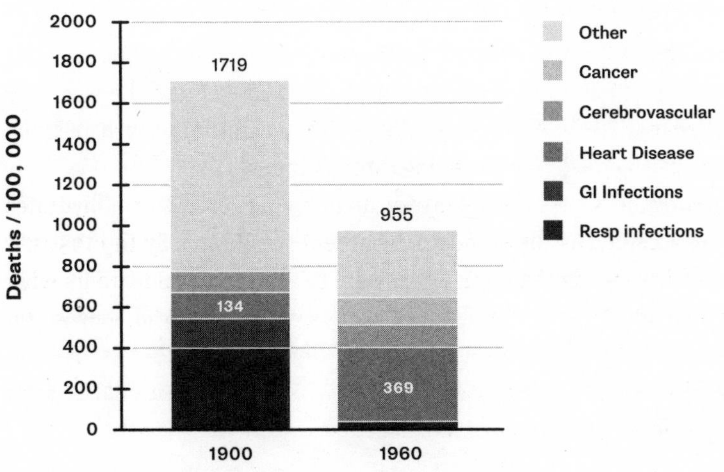

The increase in life expectancy from 1900 to 1950 reinforced the perception of a coronary-disease epidemic. For a white male, the life expectancy in 1900 was fifty years.[7] By 1950, it had reached sixty-six years, and by 1970, almost sixty-eight years. If people were not dying of tuberculosis, then they would live long enough to develop their heart attack. Currently, the average age at first heart attack is sixty-six years.[8] The risk of a heart attack in a fifty-year-old man is substantially lower than in a sixty-eight-year-old man. So the natural consequence of a longer life expectancy is an increased rate of coronary disease.

But all great stories need a villain, and dietary fat was cast into that role. Dietary fat was thought to increase the amount of cholesterol, a fatty substance that is thought to contribute to heart disease, in the blood. Soon, physicians began to advocate lower-fat diets. With great enthusiasm and shaky science, the demonization of dietary fat began in earnest.

There was a problem, though we didn't see it at the time. The three macronutrients are fat, protein and carbohydrates: lowering dietary fat meant replacing it with either protein or carbohydrates. Since many high-protein foods like meat and dairy are also high in fat, it is difficult to lower fat in the diet without lowering protein as well.

So, if one were to restrict dietary fats, then one must increase dietary carbohydrates and vice versa. In the developed world, these carbohydrates all tend to be highly refined.

Low Fat = High Carbohydrate

This dilemma created significant cognitive dissonance. Refined carbohydrates could not simultaneously be both good (because they are low in fat) and bad (because they are fattening). The solution adopted by most nutrition experts was to suggest that *carbohydrates* were no longer fattening. Instead, *calories* were fattening. Without evidence or historical precedent, it was arbitrarily decided that excess *calories* caused weight gain, not specific foods. Fat, as the dietary villain, was now deemed fattening—a previously unknown concept. The Calories-In/Calories-Out model began to displace the prevailing 'fattening carbohydrates' model.

But not everybody bought in. One of the most famous dissidents was the prominent British nutritionist John Yudkin (1910–1995). Studying diet and heart disease, he found no relationship between dietary fat and heart disease. He believed that the main culprit of both obesity and heart disease was sugar.[9,10] His 1972 book, *Pure, White and Deadly: How Sugar Is Killing Us,* is eerily prescient (and should certainly win the award for Best Book Title Ever). Scientific debate raged back and forth about whether the culprit was dietary fat or sugar.

THE DIETARY GUIDELINES

THE ISSUE WAS finally settled in 1977, not by scientific debate and discovery, but by governmental decree. George McGovern, then chairman of the United States Senate Select Committee on Nutrition and Human Needs, convened a tribunal, and after several days of deliberation, it was decided that henceforth, dietary fat was guilty as charged. Not only was dietary fat guilty of causing heart disease, but it also caused *obesity,* since fat is calorically dense.

The resulting declaration became the *Dietary Goals for the United States.* An entire nation, and soon the entire world, would now follow nutritional advice from a politician. This was a remarkable break from tradition. For the first time, a government institution intruded into the kitchens of America. Mom used to tell us what we should and should not eat. But from now on, Big Brother would be telling us. And he said, 'Eat less fat and more carbohydrates.'

Several specific dietary goals were set forth. These included

- raise consumption of carbohydrates until they constituted 55 per cent to 60 per cent of calories, and
- decrease fat consumption from approximately 40 per cent of calories to 30 per cent, of which no more than one-third should come from saturated fat.

With no scientific evidence, the formerly 'fattening' carbohydrate made a stunning transformation. While the guidelines still recognized the evils of sugar, refined grain was as innocent as a nun in a convent.

Its nutritional sins were exonerated, and it was henceforth reborn and baptized as the healthy whole grain.

Was there any evidence? It hardly mattered. The goals were now the nutritional orthodoxy. Everything else was heathen. If you didn't toe the line, you were ridiculed. The *Dietary Guidelines for Americans*, a report released in 1980 for widespread public consumption, followed the recommendations of the McGovern report closely. The nutritional landscape of the world was forever changed.

The *Dietary Guidelines for Americans*, now updated every five years, spawned the infamous food pyramid in all its counterfactual glory. The foods that formed the base of the pyramid—*the foods we should eat every single day*—were breads, pastas and potatoes. These were the precise foods that we had previously avoided to stay thin. For example, the American Heart Association's 1995 pamphlet, *The American Heart Association Diet: An Eating Plan for Healthy Americans*, declared we should eat six or more servings of 'breads, cereals, pasta and starchy vegetables (that) are low in fat and cholesterol.' To drink, 'Choose... fruit punches, carbonated soft drinks.' Ahhh. White bread and carbonated soft drinks—the dinner of champions. Thank you, American Heart Association (AHA).

Entering this brave new world, Americans tried to comply with the nutritional authorities of the day and made a conscious effort to eat less fat, less red meat, fewer eggs and more carbohydrates. When doctors advised people to stop smoking, rates dropped from 33 per cent in 1979 to 25 per cent by 1994. When doctors said to control blood pressure and cholesterol, there was a 40 per cent decline in high blood pressure and a 28 per cent decline in high cholesterol. When the AHA told us to eat more bread and drink more juice, we ate more bread and drank more juice.

Inevitably, sugar consumption increased. From 1820 to 1920, new sugar plantations in the Caribbean and American South increased the availability of sugar in the U.S. Sugar intake plateaued from 1920 to 1977. Even though 'avoid too much sugar' was an explicit goal of the

1977 *Dietary Guidelines for Americans,* consumption increased anyway until the year 2000. With all our attention focused on fat, we took our eyes off the ball. Everything was 'low fat' or 'low cholesterol,' and nobody was paying attention to sugar. Food processors, figuring this out, increased the added sugars in processed food for flavor.

Refined grain consumption increased by almost 45 per cent. Since carbohydrates in North America tended to be refined, we ate more and more low-fat bread and pasta, not cauliflower and kale.[11]

Success! From 1976 to 1996, the average fat intake decreased from 45 per cent of calories to 35 per cent. Butter consumption decreased 38 per cent. Animal protein decreased 13 per cent. Egg consumption decreased 18 per cent. Grains and sugars increased.

Until that point, the widespread adoption of the low-fat diet was completely untested. We had no idea what effect it would have on human health. But we had the fatal conceit that we were somehow smarter than 200,000 years of Mother Nature. So, turning away from the natural fats, we embraced refined low-fat carbohydrates such as bread and pasta. Ironically, the American Heart Association, even as late as the year 2000, felt that low-carbohydrate diets were dangerous fads, despite the fact that these diets had been in use almost continuously since 1863.

What was the result? The incidence of heart disease certainly did not decrease as expected. But there was definitely a consequence to this dietary manipulation—an unintentional one. Rates of obesity, defined as having a body mass index greater than 30, dramatically *increased,* starting almost exactly in 1977, as illustrated by Figure 1.2.[12]

Figure 1.2. Increase in obese and extremely obese United States adults aged 20-74.

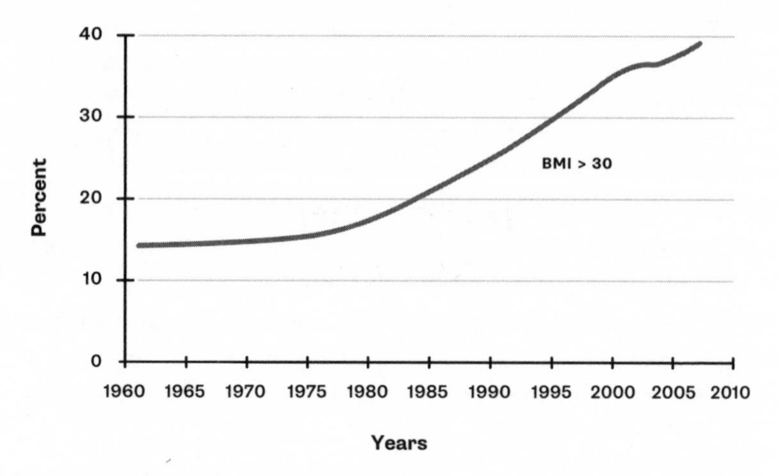

The abrupt increase in obesity began exactly with the officially sanctioned move toward a low-fat, high-carbohydrate diet. Was it mere coincidence? Perhaps the fault lay in our genetic makeup instead.

(2)

INHERITING OBESITY

•

IT IS FAIRLY obvious that obesity runs in families.[1] Obese children often have obese siblings. Obese children become obese adults.[2] Obese adults go on to have obese children. Childhood obesity is associated with a 200 per cent to 400 per cent increased risk of adult obesity. This is an undeniable fact. The controversy revolves around whether this trend is a genetic or an environmental problem—the classic nature versus nurture debate.

Families share genetic characteristics that may lead to obesity. However, obesity has become rampant only since the 1970s. Our genes could not have changed within such a short time. Genetics can explain much of the inter-individual risk of obesity, but not why entire populations become obese.

Nonetheless, families live in the same environment, eat similar foods at similar times and have similar attitudes. Families often share cars, live in the same physical space and will be exposed to the same chemicals that may cause obesity—so-called chemical obesogens. For these reasons, many consider the current environment the major cause of obesity.

Conventional calorie-based theories of obesity place the blame squarely on this 'toxic' environment that encourages eating and discourages physical exertion. Dietary and lifestyle habits have changed considerably since the 1970s including

- adoption of a low-fat, high-carbohydrate diet,
- increased number of eating opportunities per day,
- more meals eating out,
- more fast-food restaurants,
- more time spent in cars and vehicles,
- increased popularity of videos games,
- increased use of computers,
- increase in dietary sugar,
- increased use of high-fructose corn syrup and
- increased portion sizes.

Any or all of these factors may contribute to the obesogenic environment. Therefore, most modern theories of obesity discount the importance of genetic factors, believing instead that consumption of excess calories leads to obesity. Eating and moving are voluntary behaviors, after all, with little genetic input.

So—exactly how much of a role does genetics play in human obesity?

NATURE VERSUS NURTURE

THE CLASSIC METHOD for determining the relative impact of genetic versus environmental factors is to study adoptive families, thereby removing genetics from the equation. By comparing adoptees to their biological and adoptive parents, the relative contribution of environmental influences can be isolated. Dr. Albert J. Stunkard performed some of the classic genetic studies of obesity.[3] Data about biological parents is often incomplete, confidential and not easily accessible by researchers. Fortunately, Denmark has maintained a relatively complete registry of adoptions, with information on both sets of parents.

Studying a sample of 540 Danish adult adoptees, Dr. Stunkard compared them to both their adoptive and biological parents. If environmental factors were most important, then adoptees should resemble their adoptive parents. If genetic factors were most important, the adoptees should resemble their biological parents.

No relationship whatsoever was discovered between the weight of the adoptive parents and the adoptees. Whether adoptive parents were thin or fat made no difference to the eventual weight of the adopted child. The environment provided by the adoptive parents was largely irrelevant.

This finding was a considerable shock. Standard calorie-based theories blame environmental factors and human behaviors for obesity. Environmental cues such as dietary habits, fast food, junk food, candy intake, lack of exercise, number of cars, and lack of playgrounds and organized sports are believed crucial in the development of obesity. But they play virtually no role. In fact, the fattest adoptees had the thinnest adoptive parents.

Comparing adoptees to their *biological* parents yielded a considerably different result. Here there was a strong, consistent correlation between their weights. The biological parents had very little or nothing to do with raising these children, or teaching them nutritional values or attitudes toward exercise. Yet the tendency toward obesity followed them like ducklings. When you took a child away from obese parents and placed them into a 'thin' household, the child still became obese.

What was going on?

Studying identical twins raised apart is another classic strategy to distinguish environmental and genetic factors. Identical twins share identical genetic material, and fraternal twins share 25 per cent of their genes. In 1991, Dr. Stunkard examined sets of fraternal and identical twins in both conditions of being reared apart and reared together.[4] Comparison of their weights would determine the effect of the different environments. The results sent a shockwave through the obesity-research community. Approximately 70 per cent of the variance in obesity is familial.

Seventy per cent.

Seventy per cent of your tendency to gain weight is determined by your parentage. Obesity is overwhelmingly inherited.

However, it is immediately clear that inheritance cannot be the sole factor leading to the obesity epidemic. The incidence of obesity has been relatively stable through the decades. *Most of the obesity epidemic materialized within a single generation.* Our genes have not changed in that time span. How can we explain this seeming contradiction?

THE THRIFTY-GENE HYPOTHESIS

THE FIRST ATTEMPT to explain the genetic basis of obesity was the thrifty-gene hypothesis, which became popular in the 1970s. This hypothesis assumes that all humans are evolutionarily predisposed to gain weight as a survival mechanism.

The argument goes something like this: In Paleolithic times, food was scarce and difficult to obtain. Hunger is one of the most powerful and basic of human instincts. The thrifty gene compels us to eat as much as possible, and this genetic predisposition to gain weight had a survival advantage. Increasing the body's food stores (fat) permitted longer survival during times of scarce or no food. Those who tended to burn the calories instead of storing them were selectively wiped out. However, the thrifty gene is ill adapted to the modern all-you-can-eat world, as it causes weight gain and obesity. But we are simply following our genetic urge to gain fat.

Like a decomposing watermelon, this hypothesis seems quite reasonable on the surface. Cut a little deeper, and you find the rotten core. This theory has long ceased to be taken seriously. However, it is still mentioned in the media, and so its flaws bear some examination. The most obvious problem is that survival in the wild depends on not being either underweight *or* overweight. A fat animal is slower and less agile than its leaner brethren. Predators would preferentially eat the fatter prey over the harder-to-catch, lean prey. By the same token, fat predators would find it much more difficult to catch lean and swift prey. Body fatness does not always provide a survival advantage, but instead can be a significant disadvantage. How many times have you seen a fat zebra or gazelle on the *National Geographic* channel? What about fat lions and tigers?

The assumption that humans are genetically predisposed to overeat is incorrect. Just as there are hormonal signals of hunger, there are multiple hormones that tell us when we're full and *stop* us from overeating. Consider the all-you-can-eat buffet. It is impossible to simply eat and eat without stopping because we get 'full.' Continuing to eat may make us become sick and throw up. There is no genetic predisposition to overeating. There is, instead, powerful built-in protection *against* it.

The thrifty-gene hypothesis assumes chronic food shortages prevented obesity. However, many traditional societies had plentiful food year round. For example, the Tokelau, a remote tribe in the South Pacific, lived on coconut, breadfruit and fish, which were available year round. Regardless, obesity was unknown among them until the onset of industrialization and the Westernization of their traditional diet. Even in modern-day North America, widespread famine has been uncommon since the Great Depression. Yet the growth of obesity has happened only since the 1970s.

In wild animals, morbid obesity is rare, even with an abundance of food, except when it is part of the normal life cycle, as with hibernating animals. Abundant food leads to a rise in the *numbers* of animals, not an enormous increase in their *size*. Think about rats or cockroaches. When food is scarce, rat populations are low. When food is plentiful, rat populations explode. There are many more normal-sized rats, not the same number of morbidly obese rats.

There is no survival advantage to carrying a very high body-fat percentage. A male marathon runner may have 5 per cent to 11 per cent body fat. This amount provides enough energy to survive for more than a month without eating. Certain animals fatten regularly. For instance, bears routinely gain weight before hibernation—and they do so without illness. Humans, though, do not hibernate. There is an important difference between being *fat* and being *obese.* Obesity is the state of being fat to the point of having detrimental health consequences. Bears, along with whales, walruses and other fat animals are fat, but not obese, since they suffer no health consequences. They are, in fact, genetically programmed to become fat. We aren't. In humans, evolution did not favor obesity, but rather, leanness.

The thrifty-gene hypothesis doesn't explain obesity, but what does? As we will see in Part 3, 'A New Model of Obesity,' the root cause of obesity is a complex hormonal imbalance with high blood insulin as its central feature. The hormonal profile of a baby is influenced by the environment in the mother's body before birth, setting up a tendency for high insulin levels and associated obesity later in life. The explanation of obesity as a caloric imbalance simply cannot account for this predominantly genetic effect, since eating and exercise are voluntary behaviors. Obesity as a hormonal imbalance more effectively explains this genetic effect.

But inherited factors account for only 70 per cent of the tendency to obesity that we observe. The other 30 per cent of factors are under our control, but what should we do to make the most of this? Are diet and exercise the answer?

PART TWO

The Calorie Deception

(3)

THE CALORIE-REDUCTION ERROR

•

TRADITIONALLY, OBESITY HAS been seen as a result of how people process calories, that is, that a person's weight could be predicted by a simple equation:

Calories In – Calories Out = Body Fat

This key equation perpetrates what I call the calorie deception. It is dangerous precisely because it appears so simple and intuitive. But what you need to understand is that many false assumptions are built in.

Assumption 1: Calories In and Calories Out are independent of each other

This assumption is a *crucial* mistake. As we'll see later on in this chapter, experiments and experience have proven this assumption false. Caloric intake and expenditure are intimately *dependent* variables. Decreasing Calories In *triggers* a decrease in Calories Out. A 30 per cent reduction in caloric intake results in a 30 per cent decrease in caloric expenditure. The end result is minimal weight loss.

Assumption 2: Basal metabolic rate is stable

We obsess about caloric intake with barely a thought for caloric expenditure, except for exercise. Measuring caloric intake is simple, but measuring the body's total energy expenditure is complicated. Therefore, the simple but completely erroneous assumption is made that energy expenditure remains constant except for exercise. Total energy expenditure is the sum of basal metabolic rate, thermogenic effect of food, nonexercise activity thermogenesis, excess post-exercise oxygen consumption *and* exercise. The total energy expenditure can go up or down by as much as 50 per cent depending upon the caloric intake as well as other factors.

Assumption 3: We exert conscious control over Calories In

Eating is a deliberate act, so we assume that eating is a conscious decision and that hunger plays only a minor role in it. But numerous overlapping hormonal systems influence the decision of when to eat and when to stop. We consciously decide to eat in response to hunger signals that are largely hormonally mediated. We consciously stop eating when the body sends signals of satiety (fullness) that are largely hormonally mediated.

For example, the smell of frying food makes you hungry at lunchtime. However, if you have just finished a large buffet, those same smells may make you slightly queasy. The smells are the same. The decision to eat or not is principally hormonal.

Our bodies possess an intricate system guiding us to eat or not. Body-fat regulation is under automatic control, like breathing. We do not consciously remind ourselves to breathe, nor do we remind our hearts to beat. The only way to achieve such control is to have homeostatic mechanisms. Since hormones control both Calories In and Calories Out, *obesity is a hormonal, not a caloric, disorder.*

Assumption 4: Fat stores are essentially unregulated

Every single system in the body is regulated. Growth in height is regulated by growth hormone. Blood sugars are regulated by the hormones insulin and glucagon, among others. Sexual maturation is regulated by testosterone and estrogen. Body temperature is regulated by a thyroid-stimulating hormone and free thyroxine. The list is endless.

We are asked to believe, however, that growth of fat cells is essentially unregulated. The simple act of eating, without any interference from any hormones, will result in fat growth. Extra calories are dumped into fat cells like doorknobs into a sack.

This assumption has already been proven false. New hormonal pathways in the regulation of fat growth are being discovered all the time. Leptin is the best-known hormone regulating fat growth, but adiponectin, hormone-sensitive lipase, lipoprotein lipase and adipose triglyceride lipase may all play important roles. If hormones regulate fat growth, *then obesity is a hormonal, not a caloric disorder.*

Assumption 5: A calorie is a calorie

This assumption is the most dangerous of all. It's obviously true. Just like a dog is a dog or a desk is a desk. There are many different kinds of dogs and desks, but the simple statement that a dog is a dog is true. However, the real issue is this: Are all calories equally likely to cause fat gain?

'A calorie is a calorie' implies that the only important variable in weight gain is the total caloric intake, and thus, all foods can be reduced to their caloric energy. But does a calorie of olive oil cause the same metabolic response as a calorie of sugar? The answer is, obviously, no. These two foods have many easily measurable differences. Sugar will increase the blood glucose level and provoke an insulin response from the pancreas. Olive oil will not. When olive oil is absorbed by the small intestine and transported to the liver, there is no significant increase in blood glucose or insulin. The two different foods evoke vastly different metabolic and hormonal responses.

These five assumptions—the key assumptions in the caloric reduction theory of weight loss—have all been proved false. All calories are not equally likely to cause weight gain. The entire caloric obsession was a fifty-year dead end.

So we must begin again. What causes weight gain?

HOW DO WE PROCESS FOOD?

WHAT IS A calorie? A calorie is simply a unit of energy. Different foods are burned in a laboratory, and the amount of heat released is measured to determine a caloric value for that food.

All the foods we eat contain calories. Food first enters the stomach, where it is mixed with stomach acid and slowly released into the small intestine. Nutrients are extracted throughout the journey through the small and large intestines. What remains is excreted as stool.

Proteins are broken down into their building blocks, amino acids. These are used to build and repair the body's tissues, and the excess is stored. Fats are directly absorbed into the body. Carbohydrates are broken down into their building blocks, sugars. Proteins, fats and carbohydrates all provide caloric energy for the body, but differ greatly in their metabolic processing. This results in different hormonal stimuli.

CALORIC REDUCTION IS NOT THE PRIMARY FACTOR IN WEIGHT LOSS

WHY DO WE gain weight? The most common answer is that excess caloric intake causes obesity. But although the increase in obesity rates in the United States from 1971 to 2000 was associated with an increase in daily calorie consumption of roughly 200 to 300 calories,[1] it's important to remember that correlation is not causation.

Furthermore, the correlation between weight gain and the increase in calorie consumption has recently broken down.[2] Data from the National Health and Nutrition Examination Survey (NHANES) in the United States from 1990 to 2010 finds no association between increased calorie consumption and weight gain. While obesity increased at a rate of 0.37 per cent per year, caloric intake remained virtually stable. Women slightly increased their average daily intake from 1761 calories to 1781, but men slightly decreased theirs from 2616 calories to 2511.

The British obesity epidemic largely ran parallel to North America's.

But once again, the association of weight gain with increased calorie consumption does not hold true.[3] In the British experience, neither increased caloric intake nor dietary fat correlated to obesity—which argues against a causal relationship. In fact, the number of calories ingested slightly *decreased,* even as obesity rates increased. Other factors, including the nature of those calories, had changed.

We may imagine ourselves to be a calorie-weighing scale and may think that imbalance of calories over time leads to the accumulation of fat.

Calories In – Calories Out = Body Fat

If Calories Out remains stable over time, then reducing Calories In should produce weight loss. The First Law of Thermodynamics states that energy can neither be created nor destroyed in an isolated system. This law is often invoked to support the Calories In/Calories Out model. Prominent obesity researcher Dr. Jules Hirsch, quoted in a 2012 *New York Times* article,[4] explains:

> There is an inflexible law of physics—energy taken in must exactly equal the number of calories leaving the system when fat storage is unchanged. Calories leave the system when food is used to fuel the body. To lower fat content—reduce obesity—one must reduce calories taken in, or increase the output by increasing activity, or both. This is true whether calories come from pumpkins or peanuts or pâté de foie gras.

But thermodynamics, a law of physics, has minimal relevance to human biology for the simple reason that the human body is not an isolated system. Energy is constantly entering and leaving. In fact, the very act we are most concerned about—eating—puts energy into the system. Food energy is also excreted from the system in the form of stool. Having studied a full year of thermodynamics in university, I can assure you that neither calories nor weight gain were mentioned even a single time.

If we eat an extra 200 calories today, nothing prevents the body from burning that excess for heat. Or perhaps that extra 200 calories is excreted as stool. Or perhaps the liver uses the extra 200. We obsess about caloric input into the system, but output is far more important.

What determines the energy output of the system? Suppose we consume 2000 calories of chemical energy (food) in one day. What is the metabolic fate of those 2000 calories? Possibilities for their use include

- heat production,
- new protein production,
- new bone production,
- new muscle production,
- cognition (brain),
- increased heart rate,
- increased stroke volume (heart),
- exercise/physical exertion,
- detoxification (liver),
- detoxification (kidney),
- digestion (pancreas and bowels),
- breathing (lungs),
- excretion (intestines and colon) and
- fat production.

We certainly don't mind if energy is burned as heat or used to build new protein, but we *do* mind if it is deposited as fat. There are an almost infinite number of ways that the body can dissipate excess energy instead of storing it as body fat.

With the model of the calorie-balancing scale, we assume that fat gain or loss is essentially unregulated, and that weight gain and loss is under conscious control. *But no system in the body is unregulated like that.* Hormones tightly regulate every single system in the body. The thyroid, parathyroid, sympathetic, parasympathetic, respiratory, circulatory, hepatic, renal, gastrointestinal and adrenal systems are all under hormonal control. So is body fat. The body actually has multiple systems to control body weight.

The problem of fat accumulation is really a problem of *distribution* of energy. Too much energy is diverted to fat production as opposed to, say, increasing, body-heat production. The vast majority of this energy expenditure is controlled automatically, with exercise being the only factor that is under our conscious control. For example, we cannot decide how much energy to expend on fat accumulation versus new bone formation. Since these metabolic processes are virtually impossible to measure, they are assumed to remain relatively stable. In particular, Calories Out is assumed not to change in response to Calories In. We presume that the two are *independent* variables.

Let's take an analogy. Consider the money that you earn in a year (Money In) and the money that you spend (Money Out). Suppose you normally earn and also spend $100,000 per year. If Money In is now reduced to $25,000 per year, what would happen to Money Out? Would you continue to spend $100,000 per year? Probably you're not so stupid, as you'd quickly become bankrupt. Instead, you would reduce your Money Out to $25,000 per year to balance the budget. Money In and Money Out are *dependent* variables, since reduction of one will directly cause a reduction of the other.

Let's apply this reasoning to obesity. Reducing Calories In works *only* if Calories Out remains stable. What we find instead is that a sudden reduction of Calories In causes a similar reduction in Calories Out, and no weight is lost as the body balances its energy budget. Some historic experiments in calorie reduction have shown exactly this.

CALORIC REDUCTION: EXTREME EXPERIMENTS, UNEXPECTED RESULTS

EXPERIMENTALLY, IT'S EASY to study caloric reduction. We take some people, give them less to eat, watch them lose weight and live happily ever after. Bam. Case closed. Call the Nobel committee: Eat Less, Move More is the cure for obesity, and caloric reduction truly is the best way to lose weight.

Luckily for us, such studies have already been done.

A detailed study of total energy expenditure under conditions of reduced caloric intake was done in 1919 at the Carnegie Institute of Washington.[5] Volunteers consumed 'semi-starvation' diets of 1400 to 2100 calories per day, an amount calculated to be approximately 30 per cent lower than their usual intake. (Many current weight-loss diets target very similar levels of caloric intake.) The question was whether total energy expenditure (Calories Out) decreases in response to caloric reduction (Calories In). What happened?

The participants experienced a whopping 30 per cent decrease in total energy expenditure, from an initial caloric expenditure of roughly 3000 calories to approximately 1950 calories. Even nearly 100 years ago, it was clear that Calories Out is *highly dependent* on Calories In. A 30 per cent reduction in caloric intake resulted in a nearly identical 30 per cent reduction in caloric expenditure. The energy budget is balanced. The First Law of Thermodynamics is not broken.

Several decades later, in 1944 and 1945, Dr. Ancel Keys performed the most complete experiment of starvation ever done—the Minnesota Starvation Experiment, the details of which were published in 1950 in a two-volume publication entitled *The Biology of Human Starvation*.[6] In the aftermath of World War II, millions of people were on the verge of starvation. Yet the physiologic effects of starvation were virtually unknown, having never been scientifically studied. The Minnesota study was an attempt to understand both the caloric-reduction and recovery phases of starvation. Improved knowledge would help guide Europe's recovery from the brink. Indeed, as a result of this study, a relief-worker's field manual was written detailing psychological aspects of starvation.[7]

Thirty-six young, healthy, normal men were selected with an average height of five foot ten inches (1.78 meters) and an average weight of 153 pounds (69.3 kilograms). For the first three months, subjects received a standard diet of 3200 calories per day. Over the next six months of semi-starvation, only 1570 calories were given to them. However, caloric intake was continually adjusted to reach a target total weight loss of 24 per cent (compared to baseline), averaging 2.5 pounds

(1.1 kilograms) per week. Some men eventually received less than 1000 calories per day. The foods given were high in carbohydrates, similar to those available in war-torn Europe at the time—potatoes, turnips, bread and macaroni. Meat and dairy products were rarely given. In addition, they walked 22 miles per week as exercise. Following this caloric-reduction phase, their calories were gradually increased over three months of rehabilitation. Expected caloric expenditure was 3009 calories per day.[8]

Even Dr. Keys himself was shocked by the difficulty of the experiment. The men experienced profound physical and psychological changes. Among the most consistent findings was the constant feeling of cold experienced by the participants. As one explained, 'I'm cold. In July I walk downtown on a sunny day with a shirt and sweater to keep me warm. At night my well fed room mate, who isn't in the experiment, sleeps on top of his sheets but I crawl under two blankets.'[9]

Resting metabolic rate dropped by 40 per cent. Interestingly, this phenomenon is very similar to that of the previous study, which showed a drop of 30 per cent. Measurement of the subjects' strength showed a 21 per cent decrease. Heart rate slowed considerably, from an average of fifty-five beats per minute to only thirty-five. Heart stroke volume decreased by 20 per cent. Body temperature dropped to an average of 95.8°F.[10] Physical endurance dropped by half. Blood pressure dropped. Men became extremely tired and dizzy. They lost hair and their nails grew brittle.

Psychologically, there were equally devastating effects. The men experienced a complete lack of interest in everything except for food, which became an object of intense fascination to them. Some hoarded cookbooks and utensils. They were plagued with constant, unyielding hunger. Some were unable to concentrate, and several withdrew from their university studies. There were several cases of frankly neurotic behavior.

Let's reflect on what was happening here. Prior to the study, the subjects ate and also burned approximately 3000 calories per day. Then, suddenly, their caloric intake was reduced to approximately 1500 per day. All body functions that require energy experienced an immediate, across-the-board 30 per cent to 40 per cent reduction, which wrought complete havoc. Consider the following:

- Calories are needed to heat the body. Fewer calories were available, so body heat was reduced. Result: constant feeling of cold.
- Calories are needed for the heart to pump blood. Fewer calories were available, so the pump slowed down. Result: heart rate and stroke volume decreases.
- Calories are needed to maintain blood pressure. Fewer calories were available, so the body turned the pressure down. Result: blood pressure decreased.
- Calories are needed for brain function, as the brain is very metabolically active. Fewer calories were available, so cognition was reduced. Result: lethargy and inability to concentrate.
- Calories are needed to move the body. Fewer calories were available, so movement was reduced. Result: weakness during physical activity.
- Calories are needed to replace hair and nails. Fewer calories were available, so hair and nails were not replaced. Result: brittle nails and hair loss.

The body reacts in this way—by reducing energy expenditure—because the body is *smart* and doesn't want to die. What would happen if the body continued to expend 3000 calories daily while taking in only 1500? Soon fat stores would be burned, then protein stores would be burned, and then you would die. Nice. The smart course of action for the body is to immediately reduce caloric expenditure to 1500 calories per day to restore balance. Caloric expenditure may even be adjusted a little lower (say, to 1400 calories per day), to create a margin of safety. *This is exactly what the body does.*

In other words, the body *shuts down.* In order to preserve itself, it implements across-the-board reductions in energy output. The crucial point to remember is that *doing so ensures survival of the individual in a time of extreme stress.* Yeah, you might feel lousy, but you'll live to tell the tale. Reducing output is the smart thing for the body to do. Burning energy it does not have would quickly lead to death. The energy budget must be balanced.

Calories In and Calories Out are highly dependent variables.

With reflection, it should immediately be obvious that caloric expenditure *must* decrease. If we reduce daily calorie intake by 500 calories, we assume that 1 pound (0.45 kilograms) of fat per week is lost. Does that mean that in 200 weeks, we would lose 200 pounds (91 kilograms) and weigh zero pounds? Of course not. The body must, at some point, reduce its caloric expenditure to meet the lower caloric intake. It just so happens that this adaptation occurs almost immediately and persists long term. The men in the Minnesota Starvation Experiment should have lost 78 pounds (35.3 kilograms), but the actual weight lost was only 37 pounds (16.8 kilograms)—less than half of what was expected. More and more severe caloric restriction was required to continue losing weight. Sound familiar?

What happened to their weight after the semi-starvation period?

During the semi-starvation phase, body fat dropped much quicker than overall body weight as fat stores are preferentially used to power the body. Once the participants started the recovery period, they regained the weight rather quickly, in about twelve weeks. But it didn't stop there. Body weight continued to increase until it was actually higher than it was prior to the experiment.

The body quickly responds to caloric reduction by reducing metabolism (total energy expenditure), but how long does this adaptation persist? Given enough time, does the body increase its energy expenditure back to its previous higher level if caloric reduction is maintained? The short answer is no.[11] In a 2008 study, participants initially lost 10 per cent of body weight, and their total energy expenditure decreased as expected. But how long did this situation last? It remained reduced over the course of the entire study—a full year. Even after one year at the new, lower body weight, their total energy expenditure was still reduced by an average of almost 500 calories per day. In response to caloric reduction, metabolism decreases almost immediately, and that decrease persists more or less indefinitely.

The applicability of these findings to caloric-reduction diets is obvious. Assume that prior to dieting, a woman eats and burns 2000 calories per day. Following doctor's orders, she adopts a calorie-restricted, portion-controlled, low-fat diet, reducing her intake by 500 calories per day. Quickly, her total energy expenditure also drops by 500 calories per day, if not a little more. She feels lousy, tired, cold, hungry, irritable and depressed, but sticks with it, thinking that things must eventually improve. Initially, she loses weight, but as her body's caloric expenditure decreases to match her lowered intake, her weight plateaus. Her dietary compliance is good, but one year later, things have not improved. Her weight slowly creeps back up, even though she eats the same number of calories. Tired of feeling so lousy, she abandons the failed diet and resumes eating 2000 calories per day. Since her metabolism has slowed to an output of only 1500 calories per day, all her weight comes rushing back—as fat. Those around her silently accuse her of lacking willpower. Sound familiar? But her weight regain is *not* her failure. Instead, *it's to be expected.* Everything described here has been well documented over the last 100 years!

AN ERRONEOUS ASSUMPTION

LET'S CONSIDER A last analogy here. Suppose we manage a coal-fired power plant. Every day to generate energy, we receive and burn 2000 tons of coal. We also keep some coal stored in a shed, just in case we run low.

Now, all of a sudden, we receive only 1500 tons of coal a day. Should we continue to burn 2000 tons of coal daily? We would quickly burn through our stores of coal, and then our power plant would be shut down. Massive blackouts develop over the entire city. Anarchy and looting commence. Our boss tells us how utterly stupid we are and yells, 'Your ass is FIRED!' Unfortunately for us, he's entirely right.

In reality, we'd handle this situation another way. As soon as we realize that we've received only 1500 tons of coal, we'd immediately reduce our power generation to burn only 1500 tons. In fact, we might burn only 1400 tons, just in case there were further reductions in shipments. In the city, a few lights go dim, but there are no widespread blackouts. Anarchy and looting are avoided. Boss says, 'Great job. You're not as stupid as you look. Raises all around.' We maintain the lower output of 1500 tons *as long as necessary.*

The key assumption of the theory that reducing caloric intake leads to weight loss is false, since decreased caloric intake inevitably leads to decreased caloric expenditure. This sequence has been proven time and again. We just keep hoping that this strategy will somehow, this time, work. It won't. Face it. In our heart of hearts, we already know it to be true. Caloric reduction and portion-control strategies only make you tired and hungry. Worst of all... *you regain all the weight you have lost.* I know it. You know it.

We forget this inconvenient fact because our doctors, our dieticians, our government, our scientists, our politicians and our media have all been screaming at us for *decades* that weight loss is all about Calories In versus Calories Out. 'Caloric reduction is primary.' 'Eat Less, Move More.' We have heard it so often that we do not question whether it's the truth.

Instead, we believe that the fault lies in ourselves. We feel we have failed. Some silently criticize us for not adhering to the diet. Others silently think we have no willpower and offer us meaningless platitudes.

Sound familiar?

The failing isn't ours. The portion-control caloric-reduction diet is virtually *guaranteed to fail.* Eating less does not result in lasting weight loss.

EATING IS NOT UNDER CONSCIOUS CONTROL

BY THE EARLY 1990s, the Battle of the Bulge was not going well. The obesity epidemic was gathering momentum, with type 2 diabetes following closely behind. The low-fat campaign was starting to fizzle as the promised benefits had failed to materialize. Even as we were choking down our dry skinless chicken breast and rice cakes, we were getting fatter and sicker. Looking for answers, the National Institutes of Health recruited almost 50,000 post-menopausal women for the most massive, expensive, ambitious and awesome dietary study ever done. Published in 2006, this randomized controlled trial was called the Women's Health Initiative Dietary Modification Trial.[12] This trial is arguably the most important dietary study ever done.

Approximately one-third of these women received a series of eighteen education sessions, group activities, targeted message campaigns and personalized feedback over one year. Their dietary intervention was to reduce dietary fat, which was cut down to 20 per cent of daily calories. They also increased their vegetable and fruit intake to five servings per day and grains to six servings. They were encouraged to increase exercise. The control group was instructed to eat as they normally did. Those in this group were provided with a copy of the *Dietary Guidelines for Americans,* but otherwise received little help. The trial aimed to confirm the cardiovascular health and weight-reduction benefits of the low-fat diet.

The average weight of participants at the beginning of the study was 169 pounds (76.8 kilograms). The starting average body mass index was 29.1, putting participants in the overweight category (body mass index of 25 to 29.9), but bordering on obese (body mass index greater than 30). They were followed for 7.5 years to see if the doctor-recommended diet reduced obesity, heart disease and cancer as much as expected.

The group that received dietary counseling succeeded. Daily calories dropped from 1788 to 1446 a day—a reduction of 342 calories per day for over seven years. Fat as a percentage of calories decreased from

38.8 per cent to 29.8 per cent, and carbohydrates increased from 44.5 per cent to 52.7 per cent. The women increased their daily physical activity by 14 per cent. The control group continued to eat the same higher-calorie and higher-fat diet to which they were accustomed.

The results were telling. The 'Eat Less, Move More' group started out terrifically, averaging more than 4 pounds (1.8 kilograms) of weight loss over the first year. By the second year, the weight started to be regained, and by the end of the study, there was no significant difference between the two groups.

Did these women perhaps replace some of their fat with muscle? Unfortunately, the average waist circumference increased approximately 0.39 inches (0.6 centimeters), and the average waist-to-hip ratio increased from 0.82 to 0.83 inches (2.1 centimeters), which indicates these women were actually fatter than before. *Weight loss over 7.5 years of the Eat Less, Move More strategy was not even one single kilogram (2.2 pounds).*

This study was only the latest in an unbroken string of failed experiments. Caloric reduction as the primary means of weight loss has disappointed repeatedly. Reviews of the literature by the U.S. Department of Agriculture[13] highlight this failure. All these studies, of course, serve only to confirm what we already knew. Caloric reduction doesn't cause lasting weight loss. Anybody who has ever tried it can tell you.

Many people tell me, 'I don't understand. I eat less. I exercise more. But I can't seem to lose any weight.' I understand perfectly—because this advice has been *proven to fail.* Do caloric-reduction diets work? No. The Women's Health Initiative Dietary Modification Trial was the biggest, baddest, most kick-ass study of the Eat Less, Move More strategy that has even been or ever will be done—and it was a resounding repudiation of that strategy.

What is happening when we try to reduce calories and fail to lose weight? Part of the problem is the reduced metabolism that accompanies weight loss. But that's only the beginning.

HUNGER GAMES

THE CALORIES IN, Calories Out plan for weight loss assumes that we have conscious control over what we eat. But this belief ignores the extremely powerful effect of the body's hormonal state. The defining characteristic of the human body is homeostasis, or adaptation to change. Our body deals with an ever-changing environment. In response, the body makes adjustments to minimize the effects of such changes and return to its original condition. And so it is, when the body starts to lose weight.

There are two major adaptations to caloric reduction. The first change, as we have seen, is a dramatic reduction in total energy expenditure. The second key change is that the hormonal signals that stimulate hunger increase. The body is pleading with us to eat in order for it to regain the lost weight.

This effect was demonstrated in 2011, in an elegant study of hormonal adaptation to weight loss.[14] Subjects were given a diet of 500 calories per day, which produced an average weight loss of 29.7 pounds (13.5 kilograms). Next, they were prescribed a low-glycemic-index, low-fat diet for weight maintenance and were encouraged to exercise thirty minutes per day. Despite their best intentions, almost half of the weight was regained.

Various hormonal levels, including ghrelin—a hormone that, essentially, makes us hungry—were analyzed. Weight loss significantly increased ghrelin levels in the study's subjects, even after more than one year, compared to the subjects' usual baseline.

What does that mean? It means that the subjects felt hungrier and continued to feel so, right up to end of the study.

The study also measured several satiety hormones, including peptide YY, amylin and cholecystokinin, all of which are released in response to proteins and fats in our diet and serve to make us feel full. This response, in turn, produces the desired effect of keeping us from overeating. More than a year after initial weight loss, the levels of all three satiety hormones were significantly lower than before.

What does that mean? It means that the subjects felt less full.

With increased hunger and decreased satiety, the desire to eat rises. Moreover, these hormonal changes occur almost immediately and persist almost indefinitely. People on a diet tend to feel hungrier, and that effect isn't some kind of psychological voodoo, nor is it a loss of willpower. Increased hunger is a normal and expected hormonal response to weight loss.

Dr. Keys's Minnesota Starvation Experiment first documented the effect of 'semi-starvation neurosis.' People who lose weight *dream* about food. They obsess about food. All they can think about is food. Interest in all else diminishes. This behavior is not some strange affliction of the obese. In fact, it's entirely hormonally driven and normal. The body, through hunger and satiety signaling, is compelling us to get more food.

Losing weight triggers two important responses. First, total energy expenditure is immediately and indefinitely reduced in order to conserve the available energy. Second, hormonal hunger signaling is immediately and indefinitely amplified in an effort to acquire more food. Weight loss results in increased hunger and decreased metabolism. This evolutionary survival strategy has a single purpose: *to make us regain the lost weight.*

Functional magnetic resonance imaging studies show that areas of the brain controlling emotion and cognition light up in response to food stimuli. Areas of the prefrontal cortex involved with restraint show decreased activity. In other words, it is harder for people who have lost weight to resist food.[15]

This has nothing whatsoever to do with a lack of willpower or any kind of moral failure. It's a normal *hormonal fact of life.* We feel hungry, cold, tired and depressed. These are all real, measurable physical effects of calorie restriction. Reduced metabolism and the increased hunger are not the *cause* of obesity—they are the *result.* Losing weight *causes* the reduced metabolism and increased hunger, not the other way around. We do not simply make a personal choice to eat more. One of the great pillars of the caloric-reduction theory of obesity—that we eat too much because we choose to—is simply not true. We do not eat too much because we choose to, or because food is too delicious, or because of salt, sugar and fat. We eat too much because our own brain compels us to.

THE VICIOUS CYCLE OF UNDER-EATING

AND SO WE have the vicious cycle of under-eating. We start by eating less and lose some weight. As a result, our metabolism slows and hunger increases. We start to regain weight. We double our efforts by eating even less. A bit more weight comes off, but again, total energy expenditure decreases and hunger increases. We start regaining weight. So we redouble our efforts by eating even less. This cycle continues until it is intolerable. We are cold, tired, hungry and obsessing about calories. Worst of all, the weight always comes back on.

At some point, we go back to our old way of eating. Since metabolism has slowed so much, even resuming the old way of eating causes quick weight gain, up to and even a little past the original point. We are doing exactly what our hormones are influencing us to do. But friends, family and medical professionals silently blame the victim, thinking that it is 'our fault.' And we ourselves feel that we are a failure.

Sound familiar?

All dieters share this same sad story of weight loss and regain. It's a virtual guarantee. The cycle has been scientifically established, and its truth has been forged in the tears of millions of dieters. Yet nutritional authorities continue to preach that caloric reduction will lead to nirvana of permanent weight loss. In what universe do they live?

THE CRUEL HOAX

CALORIC REDUCTION IS a harsh and bitter disappointment. Yet all the 'experts' still agree that caloric reduction is the key to lasting weight loss. When you don't lose weight, they say, 'It's your fault. You were gluttons. You were sloths. You didn't try hard enough. You didn't want it badly enough.' There's a dirty little secret that nobody is willing to admit: The low-fat, low-calorie diet has already been *proven* to fail. This is the cruel hoax. Eating less does not result in lasting weight loss. It. Just. Does. Not. Work.

It is cruel because so many of us have believed it. It is cruel because all of our 'trusted health sources' tell us it is true. It is cruel because when it fails, we blame ourselves. Let me state it as plainly as I can: 'Eat Less' does not work. That's a fact. *Accept it.*

Pharmaceutical methods of caloric reduction only emphasize the spectacular failure of this paradigm. Orlistat (marketed in the U.S. as Alli) was designed to block the absorption of dietary fat. Orlistat is the drug equivalent of the low-fat, low-calorie diet.

Among its numerous side effects, the most bothersome was euphemistically called fecal leakage and oily spotting. The unabsorbed dietary fat came out the other end, where it often stained underwear. Weight-loss forums chimed in with useful advice about the 'orange poop oil.' Never wear white pants. Never assume it's just a fart. In 2007, Alli won the 'Bitter Pill Award' for worst drug from the U.S. consumer group Prescription Access Litigation. There were more serious concerns such as liver toxicity, vitamin deficiency and gallstones. However, orlistat's insurmountable problem was that it did not really work.[16]

In a randomized, double-blind controlled study,[17] four years of taking the medication three times daily resulted in an extra 6 pounds (2.8 kilograms) of weight loss. But *91 per cent* of the patients complained of side effects. It hardly seemed worth the trouble. Sales of the drug peaked in 2001 at $600 million. Despite being sold over the counter, by 2013, sales had plummeted to $100 million.

The fake fat olestra was a similarly ill-conceived notion, born out of caloric-reduction theory. Released to great fanfare several years ago, olestra was not absorbed by the body and thus had no caloric impact. Its sales began to sink within two years of release.[18] The problem? It led to no significant weight loss. By 2010, it landed on *Time* magazine's list of the fifty worst inventions, just behind asbestos.[19]

(4)

THE EXERCISE MYTH

•

DR. PETER ATTIA is the cofounder of Nutrition Science Initiative (NuSi), an organization dedicated to improving the quality of science in nutrition and obesity research. A few years ago, he was an elite long-distance swimmer, one of only a dozen or so people to have swum from Los Angeles to Catalina Island. A physician himself, he followed the standard prescribed diet high in carbohydrates and trained religiously for three to four hours daily. He was also, by his own estimation, about forty pounds (18 kilograms) overweight with a body mass index of 29 and 25 per cent body fat.

But isn't increasing exercise the key to weight loss?

Caloric imbalance—increased caloric intake combined with decreased caloric expenditure—is considered the recipe for obesity. Up until now, we've assumed that exercise was vitally important to weight loss—that by increasing exercise, we can burn off the excess calories that we eat.

THE LIMITS OF EXERCISE: A HARSH REALITY

CERTAINLY, EXERCISE HAS great health benefits. The early Greek physician Hippocrates, considered the father of medicine, said, 'If we could give every individual the right amount of nourishment and exercise, not too little and not too much, we would have found the safest way to health.' In the 1950s, along with increasing concern about heart disease, interest in physical activity and exercise began to grow. In 1955, President Eisenhower established the President's Council on Youth Fitness. By 1966, the U.S. Public Health Service began to advocate that increasing physical activity was one of the best ways to lose weight. Aerobics studios began to sprout like mushrooms after a rainstorm.

The Complete Book of Running by Jim Fixx became a runaway bestseller in 1977. The fact that he died at age fifty-two of a massive heart attack was only a minor setback to the cause. Dr. Kenneth Cooper's book *The New Aerobics* was required reading in the 1980s where I went to high school. More and more people began incorporating physical activity into their leisure time.

It seemed reasonable to expect obesity rates to fall as exercise rates increased. After all, governments around the world have poured millions of dollars into promoting exercise for weight loss, and they succeeded in getting their citizens moving. In the United Kingdom from 1997 to 2008, regular exercise increased from 32 per cent to 39 per cent in men and 21 per cent to 29 per cent in women.[1]

There's a problem, though. All this activity had no effect on obesity at all. Obesity increased relentlessly, even as we sweated to the oldies. Just take a look at Figure 4.1, on the next page.[2]

Figure 4.1. The increasing worldwide prevalence of obesity.

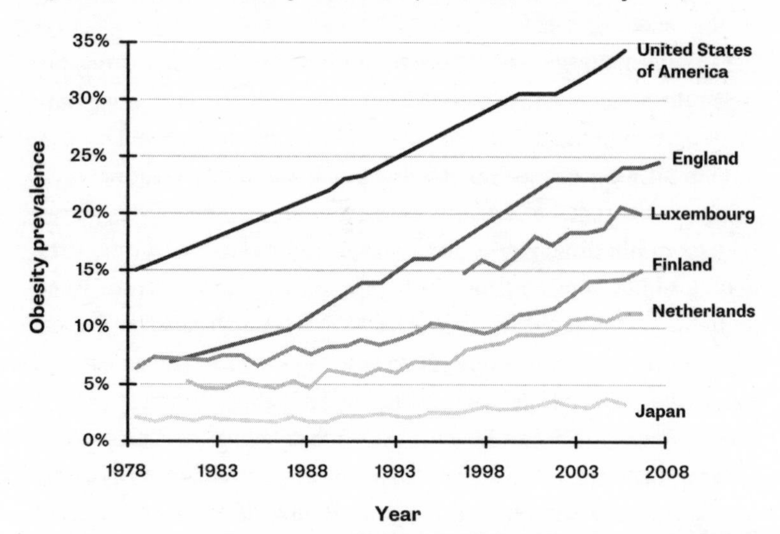

The phenomenon is global. A recent eight-country survey revealed that Americans exercised the most—135 days per year compared to a global average of 112 days. The Dutch came in last at 93 days.[3] Weight loss was the main motivation for exercise in all countries. Did all this activity translate into lower rates of obesity?

Glad you asked. The Dutch and Italians, with their low exercise rates, experienced less than one-third the obesity of those iron-pumping Americans.

The problem was apparent in the American NHANES data as well. From 2001 to 2011, there was a general increase in physical activity.[4] Certain areas (Kentucky, Virginia, Florida and the Carolinas) increased exercise at Herculean rates. But here's the dismal truth: *whether physical activity increases or decreases, it has virtually no relationship to the prevalence of obesity.* Increasing exercise did not reduce obesity. It was irrelevant. Certain states exercised more. Other states exercised less. Obesity increased by the same amount regardless.

Is exercise important in reducing childhood obesity? The short answer is no. A 2013 paper compared the physical activity (measured using accelerometry) of children aged three to five years to their weight. The authors concluded *there is no association between activity and obesity.*[5]

What went wrong?

Inherent to the Calories In, Calories Out theory is the idea that reduced physical activity plays a key role in the obesity epidemic. This idea is that we used to walk everywhere, but now we drive. With the increase in laborsaving devices such as cars, our exercise has decreased, leading to obesity. The proliferation of video games, television and computers is also believed to contribute to a sedentary lifestyle. Like any good deception, this one sounds pretty reasonable at first. There is a small problem, though. It is just not true.

Researcher Dr. Herman Pontzer studied a hunter-gatherer society living a primitive lifestyle in the modern day. The Hadza in Tanzania often travel 15 to 20 miles per day to gather food. You might assume that their daily energy expenditure is much higher than a typical office worker. Pontzer discusses the surprising results in a *New York Times* article: 'We found that despite all this physical activity, the number of calories that the Hadza burned per day was indistinguishable from that of typical adults in Europe and the United States.'[6]

Even if we compare relatively recent activity rates to those of the 1980s, before the obesity epidemic came into full swing, rates have not decreased appreciably.[7] In a Northern European population, physical-activity energy expenditure was calculated from the 1980s to the mid 2000s. The surprising finding was that if anything, physical activity has actually *increased* since the 1980s. But this study's authors went one step further. They calculated the predicted energy expenditure for a wild mammal, which is predominantly determined by body mass and ambient temperature. Compared to its wild-mammal cousins such as the seemingly vigorous cougar, fox and caribou, *Homo obesus* 2015 is *not* less physically active.

Exercise has not decreased since hunter-gatherer times, or even since the 1980s, while obesity has galloped ahead full steam. It is highly improbable that decreased exercise played any role in *causing* obesity in the first place.

If lack of exercise was not the cause of obesity epidemic, exercise is probably not going to reverse it.

CALORIES OUT

THE AMOUNT OF calories used in a day (Calories Out) is more accurately termed total energy expenditure. Total energy expenditure is the sum of basal metabolic rate (defined below), thermogenic effect of food, non-exercise activity thermogenesis, excess post-exercise oxygen consumption and, of course, exercise.

Total energy expenditure = Basal metabolic rate + Thermogenic effect of food + Nonexercise activity thermogenesis + Excess post-exercise oxygen consumption + Exercise.

The key point here is that *total energy expenditure is not the same as exercise.* The overwhelming majority of total energy expenditure is *not* exercise but the basal metabolic rate: metabolic housekeeping tasks such as breathing, maintaining body temperature, keeping the heart pumping, maintaining the vital organs, brain function, liver function, kidney function, etc.

Let's take an example. Basal metabolic rate for a lightly active average male is roughly 2500 calories per day. Walking at a moderate pace (2 miles per hour) for forty-five minutes every day, would burn roughly 104 calories. In other words, *that will not even consume 5 per cent of the total energy expenditure.* The vast majority (95 per cent) of calories are used for basal metabolism.

Basal metabolic rate depends on many factors, including

- genetics,
- gender (basal metabolic rate is generally higher in men),
- age (basal metabolic rate generally drops with age),
- weight (basal metabolic rate generally increases with muscle mass),
- height (basal metabolic rate generally increases with height),

- diet (overfeeding or underfeeding),
- body temperature,
- external temperature (heating or cooling the body) and
- organ function.

Nonexercise activity thermogenesis is the energy used in activity other than sleeping, eating or exercise; for instance, in walking, gardening, cooking, cleaning and shopping. The thermogenic effect of food is the energy used in digestion and absorption of food energy. Certain foods, such as dietary fat, are easily absorbed and take very little energy to metabolize. Proteins are harder to process and use more energy. Thermogenic effect of food varies according to meal size, meal frequency and macronutrient composition. Excess post-exercise oxygen consumption (also called after-burn) is the energy used in cellular repair, replenishment of fuel stores and other recovery activities after exercise.

Because of the complexity of measuring basal metabolic rate, nonexercise activity thermogenesis, thermogenic effect of food and excess post-exercise oxygen consumption, we make a simple but erroneous assumption that these factors are all constant over time. This assumption leads to the crucially flawed conclusion that exercise is the only variable in total energy expenditure. Thus, increasing Calories Out becomes equated with Exercise More. One major problem is that the *basal metabolic rate does not stay stable.* Decreased caloric intake can decrease basal metabolic rate by up to 40 per cent. We shall see that increased caloric intake can increase it by 50 per cent.

EXERCISE AND WEIGHT LOSS

CONVENTIONALLY, DIET AND exercise have been prescribed as treatments for obesity as if they are equally important. But diet and exercise are not fifty-fifty partners like macaroni and cheese. Diet is Batman and exercise is Robin. Diet does 95 per cent of the work and deserves all the attention; so, logically, it would be sensible to focus on diet. Exercise is still healthy and important—just not *equally* important. It has many benefits, but *weight loss is not among them.* Exercise is like brushing your teeth. It is good for you and should be done every day. Just don't expect to lose weight.

Consider this baseball analogy. Bunting is an important technique, but accounts for only perhaps 5 per cent of the game. The other 95 per cent revolves around hitting, pitching and fielding. So it would be ridiculous to spend 50 per cent of our time practicing the bunt. Or, what if we were facing a test that is 95 per cent math and 5 per cent spelling? Would we spend 50 per cent of our time studying spelling?

The fact that exercise always produces less weight loss than expected has been well documented in medical research. Studies lasting more then twenty-five weeks found that the actual weight loss was only *30 per cent of what was expected.*[8,9] In one recent controlled study, participants increased exercise to five times per week, burning 600 calories per session. Over ten months, those who exercised lost an extra ten pounds (4.5 kilograms).[10] However, the expected weight loss had been 35 pounds (16 kilograms).

Many other longer-term randomized studies have shown that exercise has minimal or no effect on weight loss.[11] A randomized 2007 study of participants who did aerobics for six days per week[12] over one year found that women reduced their weight, on average, by 3 pounds (approximately 1.4 kilograms); men, by 4 (1.8 kilograms). A Danish research team trained a previously sedentary group to run a marathon.[13] Men averaged a loss of 5 pounds (about 2.3 kilograms) of body fat. The average weight loss for women was... zero. When it comes to weight loss, exercise is just not that effective. In these cases, it was also noted that body-fat percentage was not much changed.

The Women's Health Study, the most ambitious, expensive and comprehensive diet study ever done, also looked at exercise.[14] The 39,876 women were divided into three groups representing high (more than one hour per day), medium and low levels of weekly exercise. Over the next ten years, the intense exercise group lost no extra weight. Furthermore, the study noted, 'no change in body composition was observed,' meaning that muscle was not replacing fat.

COMPENSATION: THE HIDDEN CULPRIT

WHY DOES ACTUAL weight loss fall so far below projected? The culprit is a phenomenon known as 'compensation'—and there are two major mechanisms.

First, caloric intake increases in response to exercise—we just eat more following a vigorous workout. (They don't call it 'working up an appetite' for nothing.) A prospective cohort study of 538 students from the Harvard School of Public Health found that 'although physical activity is thought of as an energy deficit activity, our estimates do not support this hypothesis.'[15] For every extra hour of exercise, the kids ate an extra 292 calories. Caloric intake and expenditure are intimately related: increasing one will cause an increase in the other. This is the biological principle of homeostasis. The body tries to maintain a stable state. Reducing Calories In reduces Calories Out. Increasing Calories Out increases Calories In.

The second mechanism of compensation relates to a reduction in non-exercise activity. If you exert yourself all day, you are less likely to exercise in your free time. The Hadza, who were walking all day, reduced their physical activity when they could. In contrast, those North Americans who were sitting all day probably increased their activity when given the chance.

This principle also holds true in children. Students aged seven and eight years who received physical education in schools were compared to those who did not.[16] The physical education group received an average of 9.2 hours per week of exercise through school, while the other group got none.

Total physical activity, measured with accelerometers, showed *there is no difference in total activity over the week between the two groups.* Why? The phys ed group compensated by doing less at home. The non-phys ed group compensated by doing more when they got home. In the end, it was a wash.

In addition, the benefit of exercise has a natural upper limit. You cannot make up for dietary indiscretions by increasing exercise. You can't outrun a poor diet. Furthermore, more exercise is not always better. Exercise represents a stress on the body. Small amounts are beneficial, but excessive amounts are detrimental.[17]

Exercise is simply not all that effective in the treatment of obesity—and the implications are enormous. Vast sums of money are spent to promote physical education in school—the Let's Move initiative, improved access to sports facilities and improved playgrounds for children—all based on the flawed notion that exercise is instrumental in the fight against obesity.

If we want to reduce obesity, we need to focus on what makes us obese. If we spend all our money, research, time and mental energy focused on exercise, we will have no resources left with which to actually fight obesity.

We are writing a final examination called Obesity 101. Diet accounts for 95 per cent of the grade and exercise for only 5 per cent. Yet we spend 50 per cent of our time and energy studying exercise. It is no wonder that our current grade is F—for Fat.

POSTSCRIPT

DR. PETER ATTIA, finally acknowledging that he was a little 'not thin,' launched a detailed self-investigation about the causes of obesity. Ignoring conventional nutritional advice and completely overhauling his diet, he was able to lose some of the excess fat that had always plagued him. The experience so moved him, that he has selflessly dedicated his career to the minefield that is obesity research.

(5)

THE OVERFEEDING PARADOX

•

SAM FELTHAM, A qualified master personal trainer, has worked in the U.K. health-and-fitness industry for more than a decade. Not accepting the caloric-reduction theory, he set out to prove it false, following the grand scientific tradition of self-experimentation. In a modern twist to the classic overeating experiments, Feltham decided that he would eat 5794 calories per day and document his weight gain. But the diet he chose was not a random 5794 calories. He followed a low-carbohydrate, high-fat diet of natural foods for twenty-one days. Feltham believed, based on clinical experience, that refined carbohydrates, not total calories, caused weight gain. The macronutrient breakdown of his diet was 10 per cent carbohydrate, 53 per cent fat and 37 per cent protein. Standard calorie calculations predicted a weight gain of about 16 pounds (7.3 kilograms). Actual weight gain, however, was only about 2.8 pounds (1.3 kilograms). Even more interesting, he dropped more than 1 inch (2.5 centimeters) from his waist measurement. He gained weight, but it was lean mass.

Perhaps Feltham was simply one of those genetic-lottery people who are able to eat anything and not gain weight. So, in the next experiment, Feltham abandoned the low-carb, high-fat diet. Instead, for twenty-one days, he ate 5793 calories per day of a standard American diet with lots of highly processed 'fake' foods. The macronutrient breakdown of his new diet was 64 per cent carbs, 22 per cent fat and 14 per cent protein—remarkably similar to the U.S. *Dietary Guidelines.* This time, the weight gain almost exactly mirrors that predicted by the calorie formula—15.6 pounds (7.1 kilograms). His waist size positively ballooned by 3.6 inches (9.14 centimeters). After only three weeks, Feltham was developing love handles.

In the same person and with an almost identical caloric intake, the two different diets produced strikingly different results. Clearly, something more than calories is at work here since diet composition apparently plays a large role. The overfeeding paradox is that excess calories alone are not sufficient for weight gain—in contradiction to the caloric-reduction theory.

OVERFEEDING EXPERIMENTS: UNEXPECTED RESULTS

THE HYPOTHESIS THAT eating too much *causes* obesity is easily testable. You simply take a group of volunteers, deliberately overfeed them and watch what happens. If the hypothesis is true, the result should be obesity.

Luckily for us, such experiments have already been done. Dr. Ethan Sims performed the most famous of these studies in the late 1960s.[1,2] He tried to force mice to gain weight. Despite ample food, the mice ate only enough to be full. After that, no inducement could get them to eat. They would not become obese. Force-feeding the mice caused an increase in their metabolism, so once again, no weight was gained. Sims then asked a devastatingly brilliant question: Could he make humans deliberately gain weight? This question, so deceptively simple, had never before been experimentally answered. After all, we already thought we knew the answer. *Of course* overfeeding would lead to obesity.

But does it really? Sims recruited lean college students at the nearby University of Vermont and encouraged them to eat whatever they wanted to gain weight. But despite what both he and the students had expected, the students could not become obese. To his utter amazement, it wasn't easy to make people gain weight after all.

While this news may sound strange, think about the last time you ate at the all-you-can-eat buffet. You were stuffed to the gills. Now can you imagine downing another two pork chops? Yeah, not so easy. Furthermore, have you ever tried to feed a baby who is absolutely refusing to eat? They scream bloody murder. It is just about impossible to make them overeat. Convincing people to overeat is not the simple task it first seems.

Dr. Sims changed course. Perhaps the difficulty here was that the students were increasing their exercise and therefore burning off the weight, which might explain their failure to gain weight. So the next step was to overfeed, but limit physical activity so that it remained constant. For this experiment, he recruited convicts at the Vermont State Prison. Attendants were present at every meal to verify that the calories—4000 per day—were eaten. Physical activity was strictly controlled.

A funny thing happened. The prisoners' weight initially rose, but then stabilized. Though at first they'd been happy to increase their caloric intake,[3] as their weight started to increase, they found it more and more difficult to overeat, and some dropped out of the study.

But some prisoners were persuaded to eat upwards of 10,000 calories per day! Over the next four to six months, the remaining prisoners did eventually gain 20 per cent to 25 per cent of their original body weight—actually much less than caloric theory predicted. Weight gain varied greatly person to person. Something was contributing to the vast differences in weight gained, but it was not caloric intake or exercise.

The key was metabolism. Total energy expenditure in the subjects *increased by 50 per cent.* Starting from an average of 1800 calories per day, total energy expenditure increased to 2700 calories per day. Their bodies tried to burn off the excess calories in order to return to their original weight. Total energy expenditure, comprising mostly basal metabolic rate, is not constant, but varies considerably in response to caloric intake. After the experiment ended, body weight quickly and effortlessly returned to normal. Most of the participants did not retain any of the weight they gained. Overeating did *not*, in fact, lead to lasting weight gain. In the same way, undereating does not lead to lasting weight loss.

In another study, Dr. Sims compared two groups of patients. He overfed a group of thin patients until they became obese. The second group was made up of very obese patients who dieted until they were only obese—but the same weight as the first group.[4] This resulted in two groups of patients who were equally heavy, but one group had originally been thin and one group originally very obese. What was the difference in total energy expenditure between the two groups? Those originally very obese subjects were burning only half as many calories as the originally thin subjects. The bodies of the originally very obese subjects were trying to return to their original higher weights by *reducing* metabolism. In contrast, the bodies of the originally thin subjects were trying to return to their original lower weights by *increasing* metabolism.

Let's return to our power plant analogy. Suppose that we receive 2000 tons of coal daily and burn 2000 tons. Now all of a sudden, we start receiving 4000 tons daily. What should we do? Say we continue to burn 2000 tons daily. The coal will pile up until all available room is used. Our boss yells, 'Why are you storing your dirty coal in my office? Your ass is FIRED!' Instead, though, we'd do the smart thing: increase coal burning to 4000 tons daily. More power is generated and no coal piles up. The boss says, 'You're doing a great job. We just broke the record for power generation. Raises all around.'

Our body also responds in a similarly smart manner. Increased caloric intake is met with increased caloric expenditure. With the increase in total energy expenditure, we have more energy, more body heat and we feel great. After the period of forced overeating, the increased metabolism quickly sheds the excess pounds of fat. The increase in nonexercise activity thermogenesis may account for up to 70 per cent of the increased energy expenditure.[5]

The results described above are by no means isolated findings. Virtually all overeating studies have produced the same result.[6] In a 1992 study, subjects were overfed calories by 50 per cent over six weeks. Body weight and fat mass did transiently increase. Average total energy expenditure increased by more than 10 per cent in an effort to burn off the excess calories. After the forced overfeeding period, body weight returned to normal and total energy expenditure decreased back to its baseline.

The paper concluded 'that there was evidence that a physiological sensor was sensitive to the fact that body weight had been perturbed and was attempting to reset it.'

More recently, Dr. Fredrik Nystrom experimentally overfed subjects double their usual daily calories on a fast-food diet.[7] On average, weight and body mass index increased 9 per cent, and body fat increased 18 per cent—by itself, no surprise. But what happened to total energy expenditure? Calories expended per day increased by 12 per cent. Even when ingesting some of the most fattening foods in the world, the body still responds to the increased caloric load by trying to burn it off.

The theory of obesity that's been dominant for the last half century—that excess calories inevitably lead to obesity—the theory that's assumed to be unassailably true, was simply not true. None of it was true.

And if excess calories don't cause weight gain, then reducing calories won't cause weight loss.

THE BODY SET WEIGHT

YOU CAN TEMPORARILY force your body weight higher than your body wants it to be by consuming excess calories. Over time, the resulting higher metabolism will reduce your weight back to normal. Similarly, you can temporarily force your body weight lower than your body wants it to be by reducing calories. Over time, the resulting lowered metabolism will raise your weight back to normal.

Since losing weight reduces total energy expenditure, many obese people assume that they have a slow metabolism, but the opposite has proved to be true.[8] Lean subjects had a mean total energy expenditure of 2404 calories, while the obese had a mean total energy expenditure of 3244 calorioes, despite spending less time exercising. The obese body was not trying to *gain* weight. It was trying to *lose* it by burning off the excess energy. So then, why are the obese... obese?

The fundamental biological principle at work here is homeostasis. There appears to be a 'set point' for body weight and fatness, as first proposed in 1984 by Keesey and Corbett.[9] Homeostatic mechanisms defend this body set weight against changes, both up and down. If weight drops below body set weight, compensatory mechanisms activate to raise it. If weight goes above body set weight, compensatory mechanisms activate to lower it.

The problem in obesity is that the set point is too high.

Let's take an example. Suppose our body set weight is 200 pounds (approximately 90 kilograms). By restricting calories, we will briefly lose weight—say down to 180 pounds (approximately 81 kilograms). If the body set weight stays at 200 pounds, the body will try to regain the lost weight by stimulating appetite. Ghrelin is increased, and the satiety hormones (amylin, peptide YY and cholecystokinin) are suppressed. At the same time, the body will decrease its total energy expenditure. Metabolism begins shutting down. Body temperature drops, heart rate drops, blood pressure drops and heart volume decreases, all in a desperate effort to conserve energy. We feel hungry, cold and tired—a scenario familiar to dieters.

Unfortunately, the result is the regain of weight back to the original body set weight of 200 pounds. This outcome, too, is familiar to dieters. Eating more is not the *cause* of weight gain but instead the *consequence*.

Eating more does not make us fat. Getting fat makes us eat more. Overeating was not a personal choice. It is a hormonally driven behavior—a natural consequence of increased hunger hormones. The question, then, is what makes us fat in the first place. In other words, *why is the body set weight so high?*

The body set weight also works in the reverse. If we overeat, we will briefly gain weight—say to 220 pounds (approximately 100 kilograms). If the body set weight stays at 200 pounds, then the body activates mechanisms to lose weight. Appetite decreases. Metabolism increases, trying to burn off the excess calories. The result is weight loss.

Our body is not a simple scale balancing Calories In and Calories Out. Rather, our body is a thermostat. The set point for weight—the body set weight—is vigorously defended against both increase and decrease. Dr. Rudolph Leibel elegantly proved this concept in 1995.[10] Subjects were deliberately overfed or underfed to reach the desired weight gain or loss. First, the group was overfed in order to gain 10 per cent of their body weight. Then, their diet was adjusted to return them to their initial weight, and then a further 10 per cent or 20 per cent weight loss was achieved. Energy expenditure was measured under all of these conditions.

As subjects' body weight increased by 10 per cent, their daily energy expenditure increased by almost 500 calories. As expected, the body responded to the intake of excess calories by trying to burn them off. As weight returned to normal, the total energy expenditure also returned to baseline. As the group lost 10 per cent and 20 per cent of their weight, their bodies reduced their daily total energy expenditure by approximately 300 calories. Underfeeding did not result in the weight loss expected because the total energy expenditure decreased to counter it. Leibel's study was revolutionary because it forced a paradigm shift in our understanding of obesity.

No wonder it is so hard to keep the weight off! Diets work well at the start, but as we lose weight, our metabolism slows. Compensatory mechanisms start almost immediately and persist almost indefinitely.

We must then reduce our caloric intake further and further simply to maintain the weight loss. If we don't, our weight plateaus and then starts to creep back up—just as every dieter already knows. (It's also hard to gain weight, but we don't usually concern ourselves with that problem, unless we are sumo wrestlers.) Virtually every dietary study of the last century has documented this finding. Now we know why.

Consider our thermostat analogy. Normal room temperature is 70°F (21°C). If the house thermostat were set instead to 32°F (0°C), we'd find it too cold. Using the First Law of Thermodynamics, we decide that the temperature of the house depends upon Heat In versus Heat Out. As fundamental law of physics, it is inviolable. Since we need more Heat In, we buy a portable heater and plug it in. But Heat In is only the proximate cause of the high temperature. The temperature at first goes up in response to the heater. But then, the thermostat, sensing the higher temperature, turns on the air conditioner. The air conditioner and the heater constantly fight against each other until the heater finally breaks. The temperature returns to 32°F.

The mistake here is to focus on the proximate and not the ultimate cause. The ultimate cause of the cold was the low setting of the thermostat. Our failure was that we did not recognize that the house contained a homeostatic mechanism (the thermostat) to return the temperature to 32°F. The smarter solution would have been for us to identify the thermostat's control and simply set it to a more comfortable 70°F and so avoid the fight between the heater and the air conditioner.

The reason diets are so hard and often unsuccessful is that we are constantly fighting our own body. As we lose weight, our body tries to bring it back up. The smarter solution is to identify the body's homeostatic mechanism and adjust it downward—and there lies our challenge. Since obesity results from a high body set weight, the treatment for obesity is to lower it. But how do we lower our thermostat? The search for answers would lead to the discovery of leptin.

LEPTIN: THE SEARCH FOR A HORMONAL REGULATOR

DR. ALFRED FROHLICH from the University of Vienna first began to unravel the neuro-hormonal basis of obesity in 1890; he described a young boy with the sudden onset of obesity who was eventually diagnosed with a lesion in the hypothalamus area of the brain. It would be later confirmed that hypothalamic damage resulted in intractable weight gain in humans.[11] This established the hypothalamic region as a key regulator of energy balance, and was also a vital clue that obesity is a hormonal imbalance.

Neurons in these hypothalamic areas were somehow responsible for setting an ideal weight, the body set weight. Brain tumors, traumatic injuries and radiation in or to this critical area cause massive obesity that is often resistant to treatment, even with a 500-calorie-per-day diet.

The hypothalamus integrates incoming signals regarding energy intake and expenditure. However, the control mechanism was still unknown. Romaine Hervey proposed in 1959 that the fat cells produced a circulating 'satiety factor.'[12] As fat stores increased, the level of this factor would also increase. This factor circulated through the blood to the hypothalamus, causing the brain to send out signals to reduce appetite or increase metabolism, thereby reducing fat stores back to normal. In this way, the body protected itself from being overweight.

The race to find this satiety factor was on.

Discovered in 1994, this factor was leptin, a protein produced by the fat cells. The name leptin was derived from 'lepto,' the Greek word for thin. The mechanism was very similar to that proposed decades earlier by Hervey. Higher levels of fat tissue produce higher levels of leptin. Traveling to the brain, it turns down hunger to prevent further fat storage.

Rare human cases of leptin deficiency were soon found. Treatment with exogenous leptin (that is, leptin manufactured outside the body) produced dramatic reversals of the associated massive obesity.

The discovery of leptin provoked tremendous excitement within the pharmaceutical and scientific communities. There was a sense that the obesity gene had, at long last, been found. However, while it played a crucial role in these rare cases of massive obesity, it was still to be determined whether it played any role in common human obesity.

Exogenous leptin was administered to patients in escalating doses,[13] and we watched with breathless anticipation as the patients... did not lose any weight. Study after study confirmed this crushing disappointment.

The vast majority of obese people are not deficient in leptin. Their leptin levels are *high*, not low. But these high levels did not produce the desired effect of lowering body fatness. Obesity is a state of *leptin resistance.*

Leptin is one of the primary hormones involved in weight regulation in the normal state. However, in obesity, it is a secondary hormone because it fails the causality test. Giving leptin doesn't make people thin. Human obesity is a disease of leptin resistance, not leptin deficiency. This leaves us with much the same question that we began with. What causes leptin resistance? What causes obesity?

PART THREE

A New Model of Obesity

(6)

A NEW HOPE

•

THE CALORIC-REDUCTION THEORY of obesity was as useful as a half-built bridge. Studies repeatedly proved it did *not* lead to permanent weight loss. Either the Eat Less, Move More strategy was ineffective, or patients were not following it. Health-care professionals could not abandon the calorie model, so what was left to do? Blame the patient, of course! Doctors and dieticians berated, ridiculed, belittled and reprimanded. They were drawn irresistibly to caloric reduction because it transformed obesity from *their* failure to understand it into *our* lack of willpower and/or laziness.

But the truth cannot be suppressed indefinitely. The caloric-reduction model was just wrong. It didn't work. Excess calories did not cause obesity, so reduced calories could not cure it. Lack of exercise did not cause obesity, so increased exercise could not cure it. The false gods of the caloric religion had been exposed as charlatans.

From those ashes, we can now begin to build a newer, more robust theory of obesity. And with greater understanding of weight gain, we have a new hope: that we can develop more rational, successful treatments.

What causes weight gain? Contending theories abound:

- Calories
- Sugar
- Refined carbohydrates
- Wheat
- All carbohydrates
- Dietary fat
- Red meat
- All meat
- Dairy products
- Snacking
- Food reward
- Food addiction
- Sleep deprivation
- Stress
- Low fiber intake
- Genetics
- Poverty
- Wealth
- Gut microbiome
- Childhood obesity

The various theories fight among themselves, as if they are all mutually exclusive and there is only one true cause of obesity. For example, recent trials that compare a low-calorie to a low-carbohydrate diet assume that if one is correct, the other is not. Most obesity research is conducted in this manner.

This approach is wrong, since these theories all contain some element of truth. Let's look at an analogy. What causes heart attacks? Consider this partial list of contributing factors:

- Family history
- Age
- Sex
- Diabetes
- Hypertension
- Hypercholesterolemia
- Smoking
- Stress
- Lack of physical activity

These factors, some modifiable and some not, all contribute to heart-attack risk. Smoking is a risk factor, but that doesn't mean that diabetes is not. All are correct since they all contribute to some degree. Nonetheless, all are also incorrect, because they are not the sole cause of heart attacks. For example, cardiovascular-disease trials would not compare smoking cessation to blood-pressure reduction since *both* are important contributing factors.

The other major problem with obesity research is that it fails to take into account that obesity is a time-dependent disease. It develops only over long periods, usually decades. A typical patient will be a little overweight as a child and slowly gain weight, averaging 1 to 2 pounds (0.5 to 1 kilogram) per year. While this amount sounds small, over forty years, the weight gained can add up to 80 pounds (35 kilograms). Given the time it takes for obesity to develop, short-term studies are of limited use.

Let's take an analogy. Suppose we were to study the development of rust in a pipe. We know that rusting is a time-dependent process that occurs over months of exposure to moisture. There would be no point in looking at studies of only one- or two-days' duration, as we might very well conclude that water does not cause pipe rust since we did not observe any rust forming during that forty-eight hours.

But this mistake is made in human obesity studies all the time. Obesity develops over decades. Yet hundreds of published studies consider only what happens in less than a year. Thousands more studies last less than a week. Still, they all claim to shed light on human obesity.

There is no clear, focused, unified theory of obesity. There is no *framework* for understanding weight gain and weight loss. This lack impedes progress in research—and so we come to our challenge: to build the *hormonal obesity theory*.

Obesity is a hormonal dysregulation of fat mass. The body maintains a body set weight, much like a thermostat in a house. When the body set weight is set too high, obesity results. If our current weight is below our body set weight, our body, by stimulating hunger and/or decreasing metabolism, will try to gain weight to reach that body set weight. Thus, excessive eating and slowed metabolism are the *result* rather than the *cause* of obesity.

But what caused our body set weight to be so high in the first place? This is, in essence, the same question as 'What causes obesity?' To find the answer, we need to know how the body set weight is regulated. How do we raise or lower our 'fat thermostat'?

THE HORMONAL THEORY OF OBESITY

OBESITY IS NOT caused by an excess of calories, but instead by a body set weight that is too high because of a hormonal imbalance in the body.

Hormones are chemical messengers that regulate many body systems and processes such as appetite, fat storage and blood sugar levels. But which hormones are responsible for obesity?

Leptin, a key regulator of body fat, did not turn out to be the main hormone involved in setting the body weight. Ghrelin (the hormone that regulates hunger) and hormones such as peptide YY and cholecystokinin that regulate satiety (feeling full or satisfied), all play a role in making you start and stop eating, but they do not appear to affect the body set weight. How do we know? A hormone suspected of causing weight gain must pass the causality test. If we inject this hormone into people, they must gain weight. These hunger and satiety hormones do not pass the causality test, but there are two hormones that do: insulin and cortisol.

In chapter 3, we saw the caloric-reduction view of obesity relies on five assumptions that have been proved to be wrong. This hormonal theory of obesity avoids making these false assumptions. Consider the following:

Assumption 1: Calories In and Calories Out are independent of each other.

The hormonal theory explains why Calories In and Calories Out are tightly synchronized with each other.

Assumption 2: Basal metabolic rate is stable.

The hormonal theory explains how hormonal signals adjust basal metabolic rate to either gain or lose weight.

Assumption 3: We exert conscious control over Calories In.

The hormonal theory explains that hunger and satiety hormones play a key role in determining whether we eat.

Assumption 4: Fat stores are essentially unregulated.
The hormonal theory explains that fat stores, like all body systems, are tightly regulated and respond to changes in food intake and activity levels.

Assumption 5: A calorie is a calorie.
The hormonal theory explains why different calories cause different metabolic responses. Sometimes calories are used to heat the body. At other times, they will be deposited as fat.

THE MECHANICS OF DIGESTION

BEFORE DISCUSSING INSULIN, we must understand hormones in general. Hormones are molecules that deliver messages to a target cell. For example, thyroid hormone delivers a message to cells in the thyroid gland to increase its activity. Insulin delivers the message to most human cells to take glucose out of the blood to use for energy.

To deliver this message, hormones must attach to the target cell by binding to receptors on the cell surface, much like a lock and key. Insulin acts on the insulin receptor to bring glucose into the cell. Insulin is the key and fits snugly into the lock (the receptor). The door opens and glucose enters. All hormones work in roughly the same fashion.

When we eat, foods are broken down in the stomach and small intestine. Proteins are broken into amino acids. Fats are broken into fatty acids. Carbohydrates, which are chains of sugars, are broken into smaller sugars. Dietary fiber is not broken down; it moves through us without being absorbed. All cells in the body can use blood sugar (glucose). Certain foods, particularly refined carbohydrates, raise blood sugar more than other foods. The rise in blood sugar stimulates insulin release.

Protein raises insulin levels as well, although its effect on blood sugars is minimal. Dietary fats, on the other hand, tend to raise both blood sugars and insulin levels minimally. Insulin is then broken down and rapidly cleared from the blood with a half-life of only two to three minutes.

Insulin is a key regulator of energy metabolism, and it is one of the fundamental hormones that promote fat accumulation and storage. Insulin facilitates the uptake of glucose into cells for energy. Without sufficient insulin, glucose builds up in the bloodstream. Type 1 diabetes results from the autoimmune destruction of the insulin-producing cells in the pancreas, which results in extremely low levels of insulin. The discovery of insulin (for which Frederick Banting and J.J.R. Macleod were awarded the 1923 Nobel Prize in Medicine), changed this formerly fatal disease into a chronic one.

At mealtimes, ingested carbohydrate leads to more glucose being available than needed. Insulin helps move this flood of glucose out of the bloodstream into storage for later use. We store this glucose by turning it into glycogen in the liver—a process is called glycogenesis. (Genesis means 'the creation of,' so this term means the creation of glycogen.) Glucose molecules are strung together in long chains to form glycogen. Insulin is the main stimulus of glycogenesis. We can convert glucose to glycogen and back again quite easily.

But the liver has only limited storage space for glycogen. Once full, excess carbohydrates will be turned into fat—a process called de novo lipogenesis. (De novo means 'from new.' Lipogenesis means 'making new fat.' De novo lipogenesis means 'to make new fat.')

Several hours after a meal, blood sugars and insulin levels start to drop. Less glucose is available for use by the muscles, the brain and other organs. The liver starts to break down glycogen into glucose to release it into general circulation for energy—the glycogen-storage process in reverse. This happens most nights, assuming you don't eat at night.

Glycogen is easily available, but in limited supply. During a short-term fast ('fast' meaning that you do not eat), your body has enough glycogen available to function. During a prolonged fast, your body can make new glucose from its fat stores—a process called gluconeogenesis (the 'making of new sugar'). Fat is burned to release energy, which is then sent out to the body—the fat-storage process in reverse.

Insulin is a storage hormone. Ample intake of food leads to insulin release. Insulin then turns on storage of sugar and fat. When there is no intake of food, insulin levels fall, and burning of sugar and fat is turned on.

This process happens every day. Normally, this well-designed, balanced system keeps itself in check. We eat, insulin goes up, and we store energy as glycogen and fat. We fast, insulin goes down and we use our stored energy. As long as our feeding and fasting periods are balanced, this system also remains balanced. If we eat breakfast at 7 a.m. and finish eating dinner at 7 p.m., the twelve hours of feeding balances the twelve hours of fasting.

Glycogen is like your wallet. Money goes in and out constantly. The wallet is easily accessible, but can only hold a limited amount of money. Fat, however, is like the money in your bank account. It is harder to access that money, but there is an unlimited storage space for energy there in your account. Like the wallet, glycogen is quickly able to provide glucose to the body. However, the supply of glycogen is limited. Like the bank account, fat stores contain an unlimited amount of energy, but they are harder to access.

This situation, of course, partially explains the difficulty in losing accumulated fat. Before getting money from the bank, you spend what's in your wallet first. But you don't like having an empty wallet. In the same manner, before getting energy from the Fat Bank, you spend the energy in the Glycogen Wallet. But you also don't like an empty Glycogen Wallet. So you keep the Glycogen Wallet filled, which prevents you from accessing the Fat Bank. In other words, before you can even begin to burn fat, you start feeling hungry and anxious because your glycogen is becoming depleted. If you continually refill your glycogen stores, you never need to use your fat stores for energy.

What happens to the excess fat that is produced through de novo lipogenesis? This newly synthesized fat can be stored as visceral fat (around organs), as subcutaneous fat (underneath the skin) or in the liver.

Under normal conditions, high insulin levels encourage sugar and fat storage. Low insulin levels encourage glycogen and fat burning. Sustained levels of excessive insulin will tend to increase fat storage. An imbalance between the feeding and fasting will lead to increased insulin, which causes increased fat, and voilà—obesity.

Could insulin be the hormonal regulator of body weight?

INSULIN, BODY SET WEIGHT AND OBESITY

OBESITY DEVELOPS WHEN the hypothalamus orders the body to increase fat mass to reach the desired body set weight. Available calories are diverted to increase fat, leaving the body short of energy (calories). The body's rational response is to try to get more calories. It increases the hormonal signals of hunger and decreases hormonal signals of satiety. We can resist the urge to eat and restrict our calorie consumption. Doing so will thwart the hypothalamus for a while, but it has other means of persuasion. The body conserves calories needed for fat growth by shutting down other functions, and metabolism slows. Increased Calories In and decreased Calories Out (eating more and moving less) does not *cause* obesity, but is instead the *result* of obesity.

Body set weight is tightly regulated. Most people's weight remains relatively stable. Even people who gain weight tend to do so extremely gradually—1 to 2 pounds per year. This does not mean, however, that body set weight is unchanging. Over time, there is a gradual upward resetting of the body's weight thermostat. The key to understanding obesity is to understand what regulates body set weight, why body set weight is set so high, and how to reset it lower.

As a key regulator of energy storage and energy balance, insulin is an obvious suspect as the body set weight regulator. If insulin causes obesity, it must do so *predominantly through its effect in the brain.* Obesity is controlled in the central nervous system through the body set weight, not in the periphery. In this hypothesis, high insulin levels increase the body set weight.

Certainly, the insulin response differs greatly between lean and obese patients. Obese patients[1] tend to have a higher fasting insulin level, as well as an exaggerated insulin response to food. (See Figure 6.1.[2]) It is possible that this hormonal activity leads to weight gain.

Figure 6.1. Different insulin responses in lean and obese people.

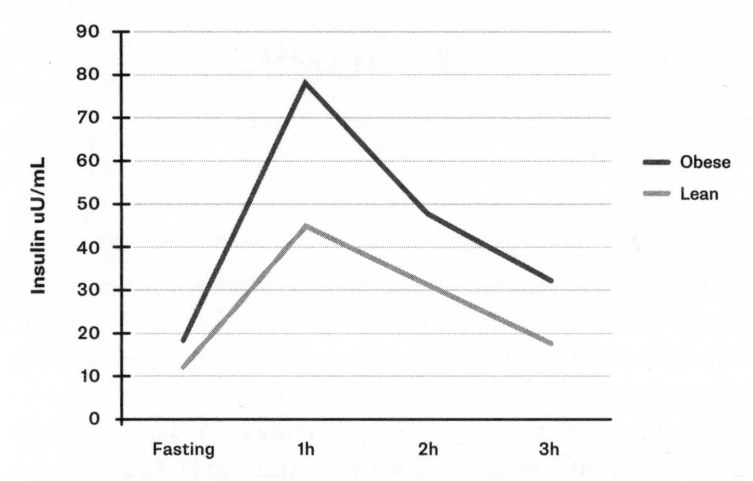

Does insulin cause obesity? That question—the key to a hormonal theory of obesity—is explored in detail in the next chapter.

(7)

INSULIN

•

I CAN MAKE YOU FAT

ACTUALLY, I CAN make anybody fat. How? By prescribing insulin. It won't matter that you have willpower, or that you exercise. It won't matter what you choose to eat. You will get fat. It's simply a matter of enough insulin and enough time.

High insulin secretion has long been associated with obesity:[1] obese people secrete much higher levels of insulin than do those of normal weight. Also, in lean subjects, insulin levels quickly return to baseline after a meal, but in the obese, these levels remain elevated.

Insulin levels are almost 20 per cent higher in obese subjects,[2] and these elevated levels are strongly correlated to important indices such as waist circumference and waist/hip ratio. The close association between insulin levels and obesity certainly suggests—but does not prove—the causal nature of this relationship.

Insulin levels can be difficult to measure since levels fluctuate widely throughout the day in response to food. It is possible to measure an 'average' level, but doing so requires multiple measurements throughout the day. Fasting insulin levels (measured after an overnight fast) are a simpler, one-step measurement. Sure enough, research reveals a close association between high fasting insulin levels and obesity, and this relationship becomes even stronger when we consider only a person's fat mass rather than his or her total weight. In the San Antonio Heart Study,[3] high fasting insulin was tightly correlated to weight gain over eight years of follow up. As we shall see in chapter 10, an insulin-resistant state leads also to high fasting insulin. This relationship is not coincidental, as insulin resistance itself plays a key role in causing obesity.

So, we know that the association between elevated insulin and obesity has already been clearly established. The question now is whether this association is, in fact, a causal relationship. Does high insulin *cause* obesity?

PUTTING IT TO THE TEST

THE 'INSULIN CAUSES obesity' hypothesis is easily tested. We can prove a causal relationship by experimentally giving insulin to a group of people and then measuring their weight gain. Therefore, for our experiment, here's our fundamental question: If you take insulin, will you get fat?

The short answer is an emphatic 'Yes!' Patients who use insulin regularly and physicians who prescribe it already know the awful truth:[4] the more insulin you give, the more obesity you get. Insulin causes obesity. Numerous studies, conducted mostly on diabetic patients, have already demonstrated this fact. Insulin causes weight gain.

Insulin is commonly used to treat both types of diabetes. In type 1 diabetes, there is destruction of the insulin-producing cells of the pancreas, resulting in very low levels of insulin. Patients require insulin injections to survive. In type 2 diabetes, cells are resistant to insulin and insulin levels are high. Patients do not always require insulin and are often treated first with oral medications.

In the landmark 1993 Diabetes Control and Complications Trial, researchers compared a standard dose of insulin to a high dose designed to tightly control blood sugars in type 1 diabetic patients.[5] At the end of six years, the study proved that intensive control of blood sugars resulted in fewer complications for those patients.

However, what happened to their weight? Participants in the high-dose group gained, on average, approximately 9.8 pounds (4.5 kilograms) more than participants in the standard group. Yowzers! More than 30 per cent of patients experienced 'major' weight gain! Prior to the study, both groups were more or less equal in weight, with little obesity. The only difference between the groups was the amount of insulin administered. Were these patients suddenly lacking in willpower? Were they lazier than they had been before the study? Were they more gluttonous? No, no and no. Insulin levels were increased. Patients gained weight.

Long-term studies in type 2 diabetes show the same weight-gaining effect of insulin.[6] The United Kingdom Prospective Diabetes Study Group, organized in the 1970s, was, at that time, the largest and longest study ever done for type 2 diabetes. Its primary purpose was to determine if intensive blood glucose management was beneficial in treating type 2 diabetes, but there were many separate sub-studies within this study. Once again, two similar groups received standard versus intensive treatment. Within the intensive group, patients were given one of two treatments—either insulin injections or a sulfonylurea drug, which increases the body's own insulin secretion. Both treatments will increase insulin levels, although by different mechanisms. Insulin injections will raise serum levels higher than the sulfonylurea.

What happened to the participants' weight? The intensive group gained an average of about 6.8 pounds (3.1 kilograms). Those that were treated with insulin gained even more—about 9 pounds (4 kilograms) on average. Increased insulin levels, whether by direct insulin injection or the use of sulfonylurea, caused significant weight gain. Once again, insulin levels were increased. Patients gained weight.

Newer types of long-acting insulin produce weight gain, too.[7] A 2007 study compared three different insulin protocols. What happened to the participants' weight? The study noted, 'Patients generally gained weight on all regimens.' Participants in the basal insulin group, which received the lowest average insulin dose, gained the least average amount of weight—4.2 pounds (1.9 kilograms). Those in the prandial insulin group, which received the most insulin, gained the most weight—12.5 pounds (5.7 kilograms) on average. The intermediate group gained on average 10.3 pounds (4.7 kilograms). The more insulin doctors gave, the more weight participants gained.

And reducing caloric intake proved useless. In a fascinating 1993 study,high-dose insulin allowed virtual normalization of blood sugars in a group of type 2 diabetic patients.[8] Starting from zero, the dose was increased to an average of 100 units per day over a period of six months. At the same time, patients decreased their caloric intake by more than 300 calories per day.

The patients' blood sugar levels were great. But what happened to their weight? It increased by an average of 19 pounds (8.7 kilograms)! Despite eating less than ever, patients gained weight like crazy. It wasn't calories that drove their weight gain. *It was insulin.*

Insulin also causes weight gain in non-diabetics. Consider what happens to patients with insulinomas—very rare insulin-secreting tumors, usually found in non-diabetics. The estimated incidence is only four cases per million per year. This tumor constantly secretes very large amounts of insulin, causing recurrent episodes of hypoglycemia (low blood sugar). But what happens to body weight? A prospective case series showed that weight gain occurs in 72 per cent of patients.[9] Removal of the tumor resulted in cure in twenty-four out of twenty-five cases. Removal of malignant insulinoma led to rapid and sustained weight loss.[10]

A 2005 case study[11] describes a twenty-year-old woman diagnosed with an insulinoma. She had gained 25 pounds over the year prior to her diagnosis. Increased caloric intake did not account for the weight gain. Reduced caloric intake did not account for the weight loss. The defining element was insulin: its rise and fall corresponded to the rise and fall in weight.

ORAL HYPOGLYCEMIC AGENTS

WE'VE SEEN THAT injections of insulin manufactured outside the body cause weight gain. There are, however, other medications, called oral hypoglycemic agents, that are taken by mouth and cause the body to produce more insulin. If these drugs also cause obesity, then that is extremely strong evidence of the *causal* link between insulin and weight gain.

Sulfonylureas and metformin

Several pills are available for the drug treatment of type 2 diabetes. The sulfonylurea class work by stimulating the pancreas to produce more insulin to lower blood sugars. All drugs in this class are well known to cause weight gain.[12]

Another oral hypoglycemic agent is metformin. Metformin decreases the amount of glucose[13] produced by the liver and increases glucose uptake by the muscles.[14]

Insulin, the sulfonylureas and metformin all have different effects on insulin levels. Insulin raises blood insulin levels the most. The sulfonylurea drug class also raises insulin levels, but not as much as insulin, and metformin does not increase insulin at all. These three treatments were compared against each other in another study.[15,16]

There was no difference in blood sugar control between the metformin group and the sulfonylurea group. But what are the effects of the different treatments on weight? Participants in the insulin group experienced the most weight gain—more than ten pounds (4.5 kilograms) on average. (We raised insulin. Patients gained weight.) Participants in the sulfonylurea group also gained weight—about 6 pounds (2.5 kilograms) on average. (We raised insulin a little. Patients gained a little weight.) Patients in the metformin group did not gain any more weight than those on diet alone. (We didn't raise insulin. Patients didn't gain weight.) *Insulin causes weight gain.*

Thiazolidinediones

The thiazolidinedione class of medications works by increasing insulin sensitivity. Thiazolidinediones do not raise insulin levels; instead, they magnify the effect of insulin, and as a result, blood sugars are lowered. Thiazolidinediones enjoyed tremendous popularity after their launch, but because of safety concerns about two of these drugs, rosiglitazone and pioglitazone, they are now rarely used.

These drugs showed a major effect other than their blood sugar–lowering ability. By amplifying insulin's effect, this insulin sensitizer caused weight gain.

Incretin agents

Incretin hormones are secreted in the stomach in response to food. These hormones may slow down stomach emptying, leading to the side effect of nausea, and also cause a short-term increase in insulin release, but only in association with meals. Several drugs that increase the effect of incretins have been tested, and in general are found to cause mild weight gain at worst, although study results vary.[17, 18] Certain incretin agents at higher doses promote weight loss, likely related to the slowing of the stomach emptying. We didn't raise insulin on a sustained basis. Weight was not gained. (Incretin agents will be discussed in much greater detail in chapter 17.)

Alpha glucosidase inhibitors

The alpha glucosidase inhibitor class of medication blocks enzymes in the small intestine that help to digest carbohydrates. As a result, the body absorbs less glucose and has lower blood glucose levels. Neither glucose use nor insulin secretion is affected.

The decrease in absorbed glucose causes a small decrease in the patient's insulin levels.[19] And what about weight? Patients had a small but statistically significant weight *loss*.[20] (We lowered insulin a little. Patients lost a little weight.)

SGLT-2 inhibitors

The newest class of medication for type 2 diabetes is the SGLT-2 (sodium-glucose linked transporter) inhibitors. These drugs block the reabsorption of glucose by the kidney, so that it spills out in the urine. This lowers blood sugars, resulting in less insulin production. SGLT-2 inhibitors can lower glucose and insulin levels after a meal by as much as 35 per cent and 43 per cent respectively.[21]

But what effect do SGLT-2 inhibitors have on weight? Studies *consistently* show a sustained and significant weight loss in patients taking these drugs.[22] Unlike virtually all dietary studies that show an initial weight loss followed by weight regain, this study found that the weight loss experienced by patients on SGLT-2 inhibitors continued for one year and longer.[23] Furthermore, their weight loss was predominantly loss of fat rather than lean muscle, although it was generally modest: around 2.5 per cent of body weight. (We lowered insulin. Patients lost weight.)

NONDIABETIC MEDICATIONS

CERTAIN MEDICATIONS UNRELATED to diabetes are also consistently related to weight gain and loss. A recent meta-analysis reviewed 257 randomized trials covering 54 different drugs to see which drugs are associated with weight change.[24]

The drug olanzapine, used to treat psychiatric disorders, is commonly associated with weight gain—5.2 pounds (2.4 kilograms) on average. Does olanzapine raise insulin levels? Absolutely—prospective studies confirm that it does.[25] As insulin rises, so does weight.

Gabapentin, a drug commonly used to treat nerve pain is also associated with weight gain, averaging 4.8 pounds (2.2 kilograms). Does it magnify insulin's effect? Absolutely. There are numerous reports of severe low blood sugars with this drug.[26] It appears that gabapentin increases the body's own insulin production.[27] Quetiapine is another antipsychotic medication associated with a smaller 2.4-pound (1.1-kilogram) average weight gain. Does it raise insulin levels? Absolutely. Insulin secretion as well as insulin resistance is increased after starting quetiapine.[28] In all these cases, we increased insulin levels. People gained weight.

I CAN MAKE YOU THIN

IF INSULIN CAUSES weight gain, can lowering its levels have the opposite effect? As insulin is reduced to very low levels, we should expect significant and severe weight loss. The SGLT-2 (sodium-glucose linked transporter) inhibitors, which lower glucose and insulin, are an example of the effect that lowering insulin may have on weight (albeit in their case, the effect is mild). Another more dramatic example is the untreated type 1 diabetic patient.

Type 1 diabetes is an autoimmune disease that destroys the insulin-producing beta cells of the pancreas. Insulin falls to extremely low levels. Blood sugar increases, but the hallmark of this condition is severe weight loss. Type 1 diabetes has been described since ancient times. Aretaeus of Cappadocia, a renowned ancient Greek physician, wrote the classic description: 'Diabetes is... a melting down of flesh and limbs into urine.' No matter how many calories the patient ingests, he or she cannot gain any weight. Until the discovery of insulin, this disease was almost universally fatal.

Insulin levels go waaayyy down. Patients lose a lot of weight.

In the type 1 diabetic community, there is a disorder called 'diabulimia.' Today, type 1 diabetic patients are treated by daily injections of insulin. There are some patients who wish to lose weight for cosmetic reasons. Diabulimia is the deliberate under-dosing of insulin for the purpose of immediate and substantial weight loss. It is extremely dangerous and certainly not advisable. However, the practice persists is because it is an extremely effective form of weight loss. Insulin levels go down. Weight is lost.

MECHANISMS

THE RESULTS ARE very consistent. Drugs that raise insulin levels cause weight gain. Drugs that have no effect on insulin levels are weight neutral. Drugs that lower insulin levels cause weight loss. The effect on weight is independent of the effect on blood sugar. A recent study suggests that 75 per cent of the weight-loss response in obesity is predicted by insulin levels.[29] Not willpower. Not caloric intake. Not peer support or peer pressure. Not exercise. Just insulin.

Insulin causes obesity—which means that insulin must be one of the major controllers of the body set weight. As insulin goes up, the body set weight goes up. The hypothalamus sends out hormonal signals to the body to gain weight. We become hungry and eat. If we deliberately restrict caloric intake, then our total energy expenditure will decrease. The result is still the same—weight gain.

As the insightful Gary Taubes wrote in his book *Why We Get Fat: And What to Do about It*, 'We do not get fat because we overeat. We overeat because we get fat.' And why do we get fat? We get fat because our body set weight thermostat is set too high. Why? Because our insulin levels are too high.

Hormones are central to understanding obesity. Everything about human metabolism, including the body set weight, is hormonally regulated. A critical physiological variable such as body fatness is not left up to the vagaries of daily caloric intake and exercise. Instead, hormones precisely and tightly regulate body fat. We don't consciously control our body weight any more than we control our heart rates, our basal metabolic rates, our body temperatures or our breathing. These are all automatically regulated, and so is our weight. Hormones tell us we are hungry (ghrelin). Hormones tell us we are full (peptide YY, cholecystokinin). Hormones increase energy expenditure (adrenalin).

Hormones shut down energy expenditure (thyroid hormone). *Obesity is a hormonal dysregulation of fat accumulation.* Calories are nothing more than the proximate cause of obesity.

Obesity is a hormonal, not a caloric imbalance.

The question of *how* insulin causes weight gain is a much more complex problem, to which all the answers are not yet known. But there are many theories.

Dr. Robert Lustig, a pediatric obesity specialist, believes that high insulin levels act as an inhibitor of leptin, the hormone that signals satiety. Leptin levels increase with body fat. This response acts on the hypothalamus in a negative feedback loop to decrease food intake and return the body to its ideal weight. However, because the brain becomes leptin resistant due to constant exposure, it does not reduce its signal to gain fat.[30]

In many ways, insulin and leptin are opposites. Insulin promotes fat storage. Leptin reduces fat storage. High levels of insulin should naturally act as an antagonist to leptin. However, the precise mechanisms by which insulin inhibits leptin are yet unknown.

Both fasting insulin and fasting leptin levels are higher in obese people, indicating a state of both insulin and leptin resistance. The leptin response to a meal was also different. In lean people, leptin levels rose—which makes sense, as leptin is a satiety hormone. However, in obese subjects, leptin levels *fell.* Despite the meal, their brains were not getting the message to stop eating. The leptin levels resistance seen in obesity may also develop due to self-regulation.[31,32] Persistently high leptin levels lead to leptin resistance. It is also possible that high insulin levels may cause increased weight gain by mechanisms unrelated to leptin in pathways yet to be discovered.

The crucial point to understand, however, is not *how* insulin causes obesity, but that insulin *does,* in fact, cause obesity.

Once we understand that obesity is a hormonal imbalance, we can begin to treat it. If we believe that excess calories cause obesity, then the treatment is to reduce calories. But this method has been a complete failure. However, if too much insulin causes obesity, then it becomes clear *we need to lower insulin levels.*

The question is *not* how to balance calories; the question is how to balance our hormones. The most crucial question in obesity is *how to reduce insulin.*

(8)

CORTISOL

•

I CAN MAKE YOU fat. Actually, I can make anybody fat. How? I prescribe prednisone, a synthetic version of the human hormone cortisol. Prednisone is used to treat many diseases, including asthma, rheumatoid arthritis, lupus, psoriasis, inflammatory bowel disease, cancer, glomerulonephritis and myasthenia gravis.

And what is one of the most consistent effects of prednisone? Like insulin, it makes you fat. Not coincidentally, both insulin and cortisol play a key role in carbohydrate metabolism. Prolonged cortisol stimulation will raise glucose levels and, subsequently, insulin. This increase in insulin plays a substantial role in the resulting weight gain.

THE STRESS HORMONE

CORTISOL IS THE so-called stress hormone, which mediates the flight-or-fight response, a set of physiological responses to perceived threats. Cortisol, part of a class of steroid hormones called glucocorticoids (glucose + cortex + steroid), is produced in the adrenal cortex. In Paleolithic times, the stress that led to a release of cortisol was often physical: for instance, being chased by a predator. Cortisol is essential in preparing our bodies for action—to fight or flee.

Once released, cortisol substantially enhances glucose availability,[1] which provides energy for muscles—very necessary in helping us to run and avoid being eaten. All available energy is directed toward surviving the stressful event. Growth, digestion and other long-term metabolic activities are temporarily restricted. Proteins are broken down and converted to glucose (gluconeogenesis).

Vigorous physical exertion (fight or flight) soon often followed, burning up these newly available stores of glucose. Shortly thereafter, we were either dead, or the danger was past and our cortisol decreased back to its normal low levels.

And that's the point: the body is well adapted to a short-term increase in cortisol and glucose levels. Over the long term, however, a problem arises.

CORTISOL RAISES INSULIN

AT FIRST GLANCE, cortisol and insulin appear have opposite effects. Insulin is a storage hormone. Under high insulin levels (mealtimes), the body stores energy in the form of glycogen and fat. Cortisol, however, prepares the body for action, moving energy out of stores and into readily available forms, such as glucose. That cortisol and insulin would have similar weight-gain effects seems remarkable—but they do. With *short-term physical stress,* insulin and cortisol play opposite roles. Something quite different happens, though, when we're under *long-term psychological stress.*

In our modern-day lives, we have many chronic, nonphysical stressors that increase our cortisol levels. For example, marital issues, problems at work, arguments with children and sleep deprivation are all serious stressors, but they do not result in the vigorous physical exertion needed to burn off the blood glucose. Under conditions of chronic stress, *glucose levels remain high* and there is no resolution to the stressor. Our blood glucose can remain elevated for months, triggering the release of insulin. Chronically elevated cortisol leads to increased insulin levels—as demonstrated by several studies.

One 1998 study showed that cortisol levels increased with self-perceived stress levels, strongly linked to increased levels of both glucose and insulin.[2] Since insulin is the major driver of obesity, it should be no surprise that both body mass index and abdominal obesity increased.

Using synthetic cortisol, we can increase insulin experimentally. Healthy volunteers given high-dose cortisol increased their insulin levels 36 per cent above their baseline.[3] Prednisone increases glucose levels by 6.5 per cent and insulin levels by 20 per cent.[4]

Over time, insulin resistance (that is, impairment of the body's ability to process insulin) also develops, mainly in the liver[5] and skeletal muscle.[6] There is a direct dose/response relationship between cortisol and insulin.[7] Long-term use of prednisone leads to an insulin-resistant state in a patient or even to full-blown diabetes.[8] This increased insulin resistance leads back to elevated insulin levels.

Glucorticoids cause muscle breakdown, releasing amino acids for gluconeogenesis, increasing blood sugars. Adiponectin, secreted by fat cells, which normally increase insulin sensitivity, are suppressed by glucocorticoids.

In a sense, insulin resistance should be expected, since cortisol generally opposes insulin. Cortisol raises blood sugar, while insulin lowers it. Insulin resistance (discussed in depth in chapter 10) is crucial to the development of obesity. Insulin resistance leads directly to higher insulin levels, and increased insulin levels are a major driver of obesity. Multiple studies confirm that increasing cortisol increases insulin resistance.[9,10,11]

If increasing cortisol raises insulin, then reducing cortisol should lower it. We find this effect in transplant patients who take prednisone (the synthetic cortisol) for years or decades as part of their anti-rejection medication. According to one study, weaning them off prednisone resulted in a 25 per cent drop in plasma insulin, which translated to a 6.0 per cent weight loss and a 7.7 per cent decrease in waist girth.[12]

CORTISOL AND OBESITY

HERE'S THE REAL question we are interested in: Does excess cortisol lead to weight gain? The ultimate test is this: Can I make somebody fat with prednisone? If so, that can prove a causal relationship, rather than a mere association. So does prednisone cause obesity? Absolutely! Weight gain is one of prednisone's most common, well-known and dreaded side effects. This relationship is causal.

It is helpful to look at what happens to people with certain diseases, particularly Cushing's disease or Cushing's syndrome, which is characterized by long-term excessive cortisol production. Cushing's disease is named for Harvey Cushing, who in 1912 described a twenty-three-year-old woman suffering from weight gain, excessive hair growth and loss of menstruation. In up to one-third of Cushing's cases, high blood sugars and overt diabetes are also present.

But the hallmark of Cushing's syndrome, even in people with mild forms, is *weight gain*. In one case series, 97 per cent of patients show abdominal weight gain and 94 per cent, increased body weight.[13, 14] Patients gain weight no matter how little they eat and no matter how much they exercise. Any disease that causes excess cortisol secretion results in weight gain. *Cortisol causes weight gain.*

However, there's evidence of the association between cortisol and weight gain even in people who don't have Cushing's syndrome. In a random sample from north Glasgow, Scotland,[15] cortisol-excretion rates were strongly correlated to body mass index and waist measurements. Higher cortisol levels were seen in heavier people. Cortisol-related weight gain, particularly abdominal fat deposits, results in an increased waist-to-hip ratio. (This effect is significant because abdominal fat deposits are more dangerous to health than all-over weight gain.)

Other measures of cortisol confirm its association with abdominal obesity. People with higher urinary cortisol excretion have higher waist-to-hip ratios.[16] People with higher cortisol in their saliva have increased body mass index and waist-to-hip ratio.[17] Long-term exposure to cortisol in the body may also be measured by scalp-hair analysis.

In a study comparing obese patients to those of normal weight, researchers found elevated levels of cortisol in scalp hair of the obese patients.[18] In other words, substantial evidence indicates that chronic cortisol stimulation increases both insulin secretion and obesity. Therefore, the hormonal theory of obesity takes shape: chronically high cortisol raises insulin levels, which in turn leads to obesity.

What about the opposite? If high cortisol levels cause weight gain, then low cortisol levels should cause weight loss. This exact situation exists in the case of Addison's disease. Thomas Addison described this classic condition, also known as adrenal insufficiency, in 1855. Cortisol is produced in the adrenal gland. When the adrenal gland is damaged, cortisol levels in the body can drop very low. The hallmark of Addison's disease is *weight loss.* Up to 97 per cent of patients exhibited weight loss.[19] (Cortisol levels went down. People lost weight.)

Cortisol may act through high insulin levels and insulin resistance, but there may also be other pathways of obesity yet to be discovered. However, the undeniable fact remains that excess cortisol causes weight gain.

And so, by extension, stress causes weight gain—something that many people have intuitively understood, despite the lack of rigorous evidence. Stress contains neither calories nor carbohydrates, but can still lead to obesity. Long-term stress leads to long-term elevated cortisol levels, which leads to extra pounds.

Reducing stress is difficult, but vitally important. Contrary to popular belief, sitting in front of the television or computer is a poor way to relieve stress. Instead, stress relief is an active process. There are many time-tested methods of stress relief, including mindfulness meditation, yoga, massage therapy and exercise. Studies on mindfulness intervention found that participants were able to use yoga, guided meditations and group discussion to successfully reduce cortisol and abdominal fat.[20]

For practical information on reducing stress through mindfulness meditation and improved sleep hygiene, see appendix C.

SLEEP

SLEEP DEPRIVATION IS a major cause of chronic stress today. Sleep duration has been steadily declining.[21] In 1910, people slept nine hours on average. However, recently, more than 30 per cent of adults between thirty and sixty-four years of age report getting fewer than six hours of sleep per night.[22] Shift workers are especially prone to sleep deprivation and often report fewer than five hours of sleep per night.[23]

Population studies consistently link short sleep duration and excess weight, generally with seven hours being the point where weight gain starts.[24,25] Sleeping five to six hours was associated with a more than 50 per cent increased risk of weight gain.[26] The more sleep deprivation, the more weight gained.

MECHANISMS

SLEEP DEPRIVATION IS a potent psychological stressor and thus stimulates cortisol. This, in turn, results in both high insulin levels and insulin resistance. A single night of sleep deprivation increases cortisol levels by more than 100 per cent.[27] By the next evening, cortisol is still 37 per cent to 45 per cent higher.[28]

Restriction of sleep to four hours in healthy volunteers resulted in a 40 per cent decrease in insulin sensitivity,[29] even after a single sleep-deprived night.[30] After five days of sleep restriction, insulin secretion increased 20 per cent and insulin sensitivity decreased by 25 per cent. Cortisol increased by 20 per cent.[31] In another study, shortened sleep duration increased the risk of type 2 diabetes.[32]

Both leptin and ghrelin, key hormones in the control of body fatness and appetite, show a daily rhythm and are disrupted by sleep disturbance. Both the Wisconsin Sleep Cohort Study and the Quebec Family study demonstrated that short sleep duration[33] is associated with higher body weight, decreased leptin and increased ghrelin.

Sleep deprivation clearly will undermine weight loss efforts.[34] Interestingly, sleep deprivation under low-stress conditions does not decrease leptin or increase hunger,[35] which suggests that it is not the sleep loss per se that is harmful, but the activation of the stress hormones and hunger mechanisms. Getting enough good sleep is essential to any weight loss plan.

(9)

THE ATKINS ONSLAUGHT

•

THE CARBOHYDRATE-INSULIN HYPOTHESIS

AS WE'VE NOW established that insulin causes obesity, our next question is: What foods causes our insulin levels to rise or to spike? The most obvious candidate is the refined carbohydrate—highly refined grains and sugars. This brings us not to a new idea, but back to a very old idea that predates even William Banting: the idea that 'fattening carbohydrates' caused obesity.

Highly refined carbohydrates are the most notorious foods for raising blood sugars. High blood sugars lead to high insulin levels. High insulin levels lead to weight gain and obesity. This chain of causes and effects has become known as the carbohydrate-insulin hypothesis. The man who found himself at the center of the controversy was the infamous Dr. Robert Atkins.

In 1963, Dr. Robert Atkins was a fat man. Like William Banting 100 years before, he needed to do something. Weighing in at 224 pounds (100 kilograms), he had recently begun his cardiology practice in New York City. He had tried the conventional ways to lose weight, but had met with no success. Recalling the medical literature published by Drs. Pennington and Gordon on low-carbohydrate diets, he decided to try the low-carbohydrate approach himself. To his amazement, it worked as advertised. Without counting calories, he shed his bothersome extra weight. He started prescribing the low-carbohydrate diet to patients and had some notable success.

In 1965, he appeared on the *Tonight Show,* and in 1970, was featured in *Vogue.* In 1972, he published his original book, *Dr. Atkins' Diet Revolution.* It was an immediate bestseller and one of the fastest-selling diet books in history.

THE LOW-CARB REVOLUTION

DR. ATKINS NEVER claimed to have invented the low-carb diet. That approach had been around long before the formerly popular diet doctor wrote about it. Jean Anthelme Brillat-Savarin wrote about carbohydrates and obesity in 1825. William Banting described the same relationship in his bestselling pamphlet, *Letter on Corpulence,* in 1863. These ideas have endured for close to two centuries.

However, by the mid 1950s, the caloric-reduction theory of obesity was gaining ascendency. It seemed so much more *scientific* to be discussing calories rather than foods. But there were still holdouts. Dr. Alfred Pennington wrote an editorial in the *New England Journal of Medicine* in 1953 emphasizing the role of carbohydrates in obesity.[1] Studies by Dr. Walter Bloom comparing low-carbohydrate diets to fasting regimens had found comparable weight loss between the two.[2]

Dr. Irwin Stillman wrote *The Doctor's Quick Weight Loss Diet* in 1967, recommending a high-protein, low-carbohydrate diet.[3] It quickly sold more than 2.5 million copies. Since it takes extra energy to metabolize dietary protein (the thermogenic effect of food), eating more protein could theoretically cause more weight loss. Dr. Stillman himself lost fifty pounds following the 'Stillman diet,' which contained up to 90 per cent protein. He reportedly used the diet to treat more than

10,000 overweight patients. By the time Dr. Atkins joined the fray, the low-carbohydrate revolution was already well underway.

Dr. Atkins argued in his 1972 bestseller that severely restricting carbohydrates would keep insulin levels low, thus reducing hunger and eventually leading to weight loss. It didn't take long for the nutritional authorities to respond. In 1973, the American Medical Association's Council on Foods and Nutrition published a blistering attack on Atkins's ideas. Most physicians at that time worried that the high fat content of the diet would lead to heart attacks and strokes.[4]

Nonetheless, low-carb proponents continued to preach. In 1983, Dr. Richard Bernstein, himself a type 1 diabetic since age nine, opened a controversial clinic to treat diabetics with a strict low-carbohydrate diet—a method that directly contradicted most nutritional and medical teachings of the time. In 1997, Bernstein published *Dr. Bernstein's Diabetes Solution.* In 1992 and then again in 1999, Atkins updated his bestseller with the publication of *Dr. Atkins' New Diet Revolution.* Bernstein's and Atkins's books would become monster bestsellers, with more than 10 million copies sold. In 1993, scientists Rachael and Richard Heller wrote *The Carbohydrate Addict's Diet,* which sold more than 6 million copies. The Atkins onslaught had well and truly begun.

The low-carb diet's popularity, rekindled in the 1990s, ignited into a full-scale inferno in 2002 when award-winning journalist Gary Taubes wrote a controversial lead article in the *New York Times* entitled 'What If It's All Been a Big Fat Lie?' He argued that dietary fat, long believed to cause atherosclerosis, was actually quite harmless to human health. He followed that up with the best-selling books *Good Calories, Bad Calories* and *Why We Get Fat,* in which he expounded on the idea that carbohydrates were the root cause of weight gain.

THE EMPIRE STRIKES BACK

THESE IDEAS WERE slow to take hold in the medical community. Many physicians still felt that low-carb was simply the latest in a long line of failed dietary fads. The American Heart Association (AHA) published its own book called the *No-Fad Diet: A Personal Plan for Healthy Weight Loss.* It's only mildly ironic that while condemning other diets, the AHA would recommend the only diet (low-fat) repeatedly proven to fail. But the low-fat religion was enshrined in the medical community and it did not tolerate disbelievers. Despite a stunning lack of evidence to support this low-fat advice, medical associations such as the AHA and the American Medical Association were quick to defend their beliefs and denounce these new 'fad' diets. But the Atkins onslaught was relentless. In 2004, more than 26 million Americans claimed to be on some type of low-carbohydrate diet. Even fast-food chains introduced low-carb lettuce-wrapped burgers. The possibility of permanently reducing excess weight and all its associated health complications seemed within grasp.

The AHA admitted that the reduced-fat diet was unproven over the long term. It also conceded that the Atkins diet evidenced a superior cholesterol profile and yielded a more rapid initial weight loss. Despite these benefits, the AHA maintained its concerns with atherogenicity—the rate at which plaques would form in the arteries. There was, of course, no evidence to support this concern. Regarding its own recommended but scientifically unsupported low-fat diet, the AHA had no concerns at all!

No concern that higher intake of sugar and other refined carbohydrates could be harmful. No concern that the low-fat diet had been proved a spectacular failure by every dietary study done. No concern that the obesity and diabetes epidemics were raging full force under their very noses. The AHA fiddled while Rome burned.

During the forty years that the AHA advised a low-fat diet, the obesity crisis grew to gargantuan proportions. Yet at no time did the AHA question whether their completely ineffectual advice was actually helping people. Instead, doctors played their favorite game: blame the patient. It is not *our* fault the diet doesn't work. It is *their* fault for not following the diet.

LOW-CARB DIETS: A STUNNED MEDICAL COMMUNITY

AS THE NEW competitor challenged conventional dietary wisdom, the campaign of slurs and innuendo began. Nonetheless, new studies started appearing by the mid 2000s comparing the 'new' low-carb diets to the old standards. The results would shock many, myself included. The first study, published in the prestigious *New England Journal of Medicine* in 2003,confirmed greater short-term weight loss with the Atkins diet.[5] In 2007, the *Journal of the American Medical Association* published a more detailed study.[6] Four different popular weight plans were compared in a head-to-head trial. One clear winner emerged—the Atkins diet. The other three diets (Ornish, which has very low fat; the Zone, which balances protein, carbohydrates and fat in a 30:40:30 ratio; and a standard low-fat diet) were fairly similar with regard to weight loss. However, in comparing the Atkins to the Ornish, it became clear that not only was weight loss better, but so was the entire metabolic profile. Blood pressure, cholesterol and blood sugars all improved to a greater extent on Dr. Atkins's diet.

In 2008, the DIRECT (Dietary Intervention Randomized Controlled Trial) study reaffirmed once again the superior short-term weight reduction of the Atkins diet.[7] Done in Israel, it compared the Mediterranean, the low-fat and the Atkins diets. While the Mediterranean diet held its own against the powerful, fat-reducing Atkins diet, the low-fat AHA standard was left choking in the dust—sad, tired and unloved, except by academic physicians. More importantly, the metabolic benefits of both the Atkins and Mediterranean diets were confirmed. The Atkins diet reduced average blood sugar levels by 0.9 per cent, far more than the other diets and almost as powerful as most medications.

The high-protein, low-glycemic index diet maintained weight loss better than the low-fat diet over six months.[8] Part of the reason may be that different weight-loss diets provoke different changes in total energy expenditure. Dr. David Ludwig from Harvard University found that the low-fat diet slowed body metabolism the most.[9] What was the best diet for maintaining metabolism? The very-low-carbohydrate diet. This diet also seemed to reduce appetite. Dr. G. Boden wrote in the *Annals of Internal Medicine* in 2005, 'When we took away the carbohydrates, the patients spontaneously reduced their daily energy consumption by 1,000 calories a day.'[10] Insulin levels dropped and insulin sensitivity was restored.

Perhaps eating refined carbohydrates leads to 'food addictions.' Natural satiety signals are hormones that are extremely powerful deterrents to overeating. Hormones such as cholecystokinin and peptide YY respond to ingested proteins and fats to signal us to stop eating. Now, let's return to that all-you-can-eat buffet mentioned in chapter 5. At some point, you simply cannot eat any more, and the idea of consuming two more pork chops is sickening. That feeling is your satiety hormones telling you that you've had enough.

But what if you were offered a small slice of cake or apple pie? Doesn't seem so hard to eat now, does it? As kids, we used to call this the second-stomach phenomenon: after the first stomach for regular food was full, we imagined that there was a second one for desserts. Somehow, despite being full, we still have room for highly refined carbohydrates like cake and pie—but not proteins or fats. Highly refined and processed foods somehow do not trigger the release of satiety hormones, and we go ahead and eat that cake.

Think about foods that people say they're 'addicted' to. Pasta, bread, cookies, chocolate, chips. Notice anything? All are highly refined carbohydrates. Does anybody ever say they are addicted to fish? Apples? Beef? Spinach? Not likely. Those are all delicious foods, but not addictive.

Consider some typical comfort foods. Macaroni and cheese. Pasta. Ice cream. Apple pie. Mashed potatoes. Pancakes. Notice anything? All are highly refined carbohydrates. There is evidence that these foods activate the reward systems in our brains, which gives us 'comfort.' Refined carbohydrates are easy to become addicted to and overeat precisely because there are no natural satiety hormones for refined carbs. The reason, of course, is that refined carbohydrates are not natural foods but are instead highly processed. *Their toxicity lies in that processing.*

THE ATKINS DECLINE

THE STUDIES MENTIONED above left the medical profession stunned and a little bit flabbergasted. Each had been undertaken almost with the express purpose of destroying the Atkins reputation. They came to bury the Atkins diet, but instead had crowned it. One by one, the concerns of the low-carb movement were put to rest. The New Diet Revolution was on pace. Long live the Revolution. But trouble was on the horizon.

Longer-term studies of the Atkins diet failed to confirm the much hoped-for benefits. Dr. Gary Foster from Temple University published two-year results showing that both the low-fat and the Atkins groups had lost but then regained weight at virtually the same rate.[11] After twelve months, all the DIRECT study patients, including the Atkins group, regained much of the weight they'd lost.[12] A systematic review of all the dietary trials showed that much of the benefits of a low-carbohydrate approach evaporated after one year.[13]

Greater compliance was supposed to be one of the main benefits of the Atkins approach, since there was no need for calorie counting. However, following the severe food restrictions of Atkins proved no easier for dieters than conventional calorie counting. Compliance was equally low in both groups, with upwards of 40 per cent abandoning the diet within one year.

In hindsight, this outcome was somewhat predictable. The Atkins diet severely restricted highly indulgent foods such as cakes, cookies, ice cream and other desserts. These foods are clearly fattening, no matter what diet you believe in. We continue to eat them simply because they *are* indulgent. Food is a celebration, and feasting has accompanied celebration throughout human history. This is as true in year 2015 AD as it was in year 2015 BC. Birthdays, weddings and holiday celebrations—what do we eat? Cake. Ice cream. Pie. Not whey powder shakes and lean pork. Why? Because we *want* to indulge. The Atkins diet does not allow for this simple fact, and that doomed it to failure.

The first-hand experience of many people confirmed that the Atkins diet was not a lasting one. Millions of people abandoned the Atkins approach, and the New Diet Revolution faded into just another dietary fad. The company Atkins Nutritionals, founded in 1989 by Dr. Atkins, filed for bankruptcy, having sustained heavy losses as its customers fled. The weight-loss benefits could not be sustained.

But why? What happened? One of the founding principles of the low-carbohydrate approach is that dietary carbohydrates increase blood sugars the most. High blood sugars lead to high insulin. High insulin is the key driver of obesity. Those facts seem reasonable enough. What was wrong?

THE CARBOHYDRATE-INSULIN HYPOTHESIS WAS INCOMPLETE

THE CARBOHYDRATE-INSULIN HYPOTHESIS, the idea that carbohydrates cause weight gain because of insulin secretion, was not exactly wrong. Carbohydrate-rich foods certainly do increase insulin levels to a greater extent than the other macronutrients. High insulin certainly does lead to obesity.

However, the hypothesis stands incomplete. There are many problems, with the paradox of the Asian rice eater being the most obvious. Most Asians, for at least the last half-century, ate a diet based on white, polished rice, a highly refined carbohydrate. Yet until recently, obesity remained quite rare in these populations.

The International Study of Macronutrients and Blood Pressure (INTERMAP) compared the diets of the U.S., U.K., China and Japan in detail[14] (see Figure 9.1[15]). This study was done in the late 1990s before globalization westernized the Asian diet.

Figure 9.1. The intermap study (2003) found that although people in China and Japan had high intakes of carbohydrates, sugar intake was lower in these countries than in the U.S. and U.K.

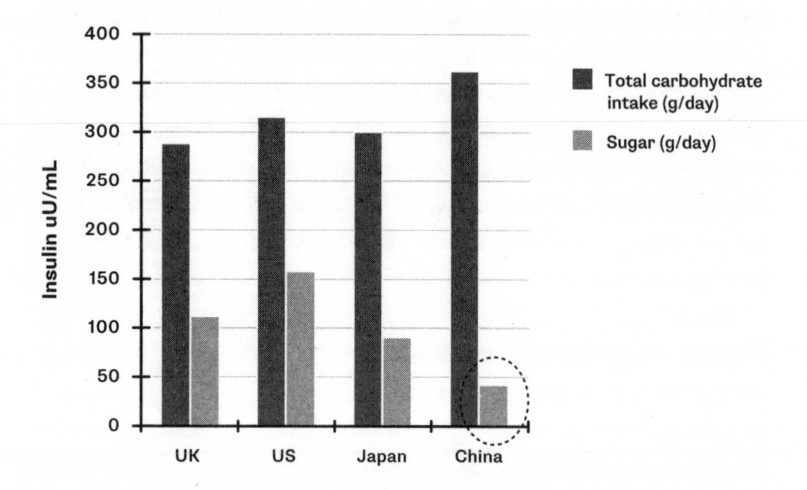

Total and percentage carbohydrate intake in China far exceeds the other nations. Sugar intake in China, however, is extremely low compared to the other nations. Japan's carbohydrate intake is similar to that of the U.K. and the U.S., but its sugar consumption is far lower. Despite high carbohydrate intakes, obesity rates in China and Japan stayed very low until recently.

So the carbohydrate-insulin hypothesis was not incorrect, but clearly something else was going on. Total carbohydrate intake was not the entire story. Sugar seemed to be contributing much more to obesity than other refined carbohydrates.

Indeed, many primitive societies that eat mostly carbohydrates have very low obesity rates. In 1989, Dr. Staffan Lindeberg studied the residents of Kitava, one of the Trobriand Islands in Papua New Guinea's archipelago—one of the last places on Earth where people ate a largely traditional diet. Starchy vegetables, including yam, sweet potato, taro and cassava, made up the basis of their diet. An estimated 69 per cent of calories were derived from carbohydrates, and less than 1 per cent of the calories came from processed Western foods. Despite this high carbohydrate intake, insulin was very low among the Kitavans, resulting in virtually no obesity. Comparing the Kitavans to his native Swedish population, Dr. Lindeberg found that despite a diet that was 70 per cent carbohydrate (unrefined), the Kitavans had insulin levels below the 5th percentile of the Swedes.[16] The average Kitavan native had an insulin level lower than 95 *per cent* of Swedes. The body mass index of young Kitavans averaged 22 (normal) and it *decreased* with age. The possibility that increased exercise led to low insulin levels and less obesity was investigated but this turned out not to be the case.

Similarly, natives of the Japanese island of Okinawa eat a diet that is nearly 85 per cent unrefined carbohydrates. The dietary staple is sweet potato. They eat three times as many green and yellow vegetables, but only 25 per cent of the sugar consumed by residents of nearby Japan. Despite the high intake of carbohydrates, there is virtually no obesity, and the average body mass index is only 20.4. They are one of the longest-lived peoples in the world, with more than triple the rate (compared to nearby Japan) of people living past 100 years.

Clearly, the carbohydrate-insulin hypothesis is an incomplete theory, leading many to abandon it rather than try to reconcile it with the known facts. One possibility is that there is an important difference in eating rice versus wheat. Asians tend to eat rice, whereas Western societies tend to take their carbohydrate as refined wheat and corn products. It is also possible that changes in Western obesity rates are related to changes in the variety of wheat we are eating. Dr. William Davis, author of *Wheat Belly*, a *New York Times* bestseller, suggests that the dwarf wheat that we eat today may be far different from the original wheat. The Einkorn variety of wheat has been cultivated since 3300 BC. By the 1960s, as the world's population grew larger, agricultural techniques aimed at increasing the yield of the wheat led to new varieties of wheat called dwarf and semi-dwarf wheat. Currently, 99 per cent of commercially grown wheat is dwarf and semi-dwarf varieties, and it may be that there are health implications of eating these new varieties of wheat.

Insulin and obesity are still causally linked. However, it is not at all clear that high carbohydrate intake is always the primary cause of high insulin levels. In Kitava, high carbohydrate intake did not lead to elevated insulin. The notion that carbohydrates are the *only* driver of insulin is incorrect. A critical piece of the puzzle had been neglected. Specifically, sugar plays a crucial role in obesity, but how does it fit in? The missing link was insulin resistance.

(10)

INSULIN RESISTANCE: THE MAJOR PLAYER

•

OPRAH WINFREY HAS waged her weight loss battles publicly for several decades. At her heaviest, she weighed 237 pounds (107.5 kilograms). By 2005, she'd battled her way to a relatively svelte 160 pounds (72.6 kilograms). She was exultant. She'd cut her carbohydrates. She'd exercised. She had a personal chef and a personal trainer. She did everything 'right.' She had every advantage not available to the rest of us. So why did she gain back 40 pounds (18 kilograms) by 2009? Why couldn't she keep the weight off?

Why is long-standing obesity so difficult to treat?

Time dependence in obesity is almost universally understood but rarely acknowledged. Usually, obesity is a gradual process of gaining 1 to 2 pounds (0.5 to 1 kilogram) per year. Over a period of twenty-five years, though, that can add up to 50 extra pounds (23 kilograms). Those who have been obese their entire lives find it extremely difficult to lose weight. In contrast, people with recent weight gain have a much, much easier time dropping the excess pounds.

Conventional caloric theories of obesity assume that losing 10 pounds (4.5 kilograms) is the same experience whether you've been overweight for one week or one decade. If you reduce the calories, the weight will be lost. But this is simply not true. Likewise, the carbohydrate-insulin hypothesis makes no allowance for duration of obesity: reducing carbohydrates should cause weight loss, regardless of how long you've been overweight. But that's not true either.

But the time frame matters *a lot*. We may try to downplay its effects, but the idea that long-standing obesity is much more difficult to treat has the stench of truth.

So we must acknowledge the phenomenon of time dependence. Obesity at age seventeen has consequences that reach *decades* into the future.[1] Any comprehensive theory of obesity must be able to explain why its duration matters *so much*.

High insulin levels cause weight gain. Food choices play a role in raising insulin levels. But we are missing yet another pathway that increases insulin, one that is both time dependent and independent of diet: *insulin resistance*.

Insulin resistance is Lex Luthor. It is the hidden force behind most of modern medicine's archenemies, including obesity, diabetes, fatty liver, Alzheimer's disease, heart disease, cancer, high blood pressure and high cholesterol. But while Lex Luthor is fictional, the insulin resistance syndrome, also called the metabolic syndrome, is not.

HOW DO WE DEVELOP RESISTANCE?

THE HUMAN BODY is characterized by the fundamental biological principle of homeostasis. If things change in one direction, the body reacts by changing in the opposite direction to return closer to its original state. For instance, if we become very cold, the body adapts by increasing body-heat generation. If we become very hot, the body sweats to try to cool itself. Adaptability is a prerequisite for survival and generally holds true for all biological systems. In other words, the body develops resistance. The body resists change out of its comfort range by adapting to it.

What happens in the case of insulin resistance? As discussed before, a hormone acts on a cell as a key that fits into a lock. When insulin (the key) no longer fits into the receptor (the lock), the cell is called insulin resistant. Because the fit is poor, the door does not open fully. As a result, less glucose enters. The cell senses that there is too little glucose inside. Instead, glucose is piling up outside the door. Starved for glucose, the cell demands more. To compensate, the body produces extra keys (insulin). The fit is still poor, but more doors are opened, allowing a normal amount of glucose to enter.

Suppose that in the normal situation we produce ten keys (insulin). Each key opens a locked door that lets two glucose molecules inside. With ten keys, twenty glucose molecules enter the cell. Under conditions of resistance, the key does not fully open the locked door. Only one glucose molecule is allowed in. With ten keys, only ten glucose molecules are allowed in. To compensate, we now produce a total of twenty keys. Now, twenty glucose molecules are allowed in, but only because we have increased the number of keys. As we develop insulin resistance, our bodies increase our insulin levels to get the same result—glucose in the cell. However, we pay the price in constantly elevated insulin levels.

Why do we care? *Because insulin resistance leads to high insulin levels,* and as we've seen, high insulin levels cause obesity.

But what caused the insulin resistance in the first place? Does the problem lie with the key (insulin) or the lock (insulin receptor)? Insulin is the same hormone, whether found in an obese or a lean person. There is no difference in amino-acid sequence or any other measurable quality. Therefore, the problem of insulin resistance must lie with the receptor. The insulin receptor does not respond properly and locks the glucose out of the cell. But why?

To begin solving this puzzle, let us back up and look for clues from other biological systems. There are many examples of biological resistance. While they may not apply specifically to the insulin/insulin-receptor problem, they may shed some light on the problem of resistance and show us where to begin.

ANTIBIOTIC RESISTANCE

LET'S START WITH antibiotic resistance. When new antibiotics are introduced, they kill virtually all the bacteria they're designed to kill. Over time, some bacteria develop the ability to survive high doses of these antibiotics. They've become drug-resistant 'superbugs,' and infections from them are difficult to treat and can sometimes lead to death. Superbug infections are a large and growing problem in many urban hospitals worldwide. All antibiotics have begun to lose their effectiveness due to resistance.

Antibiotic resistance is not new. Alexander Fleming discovered penicillin in 1928. Mass production of it was perfected by 1942, with funds from the U.S. and British governments for use in World War II. In his 1945 Nobel lecture, 'Penicillin,' Dr. Fleming correctly predicted the emergence of resistance. He said,

> There is the danger that the ignorant man may easily underdose himself and by exposing his microbes to non lethal quantities of the drug make them resistant. Here is a hypothetical illustration. Mr. X. has a sore throat. He buys some penicillin and gives himself, not enough to kill the streptococci but enough to educate them to resist penicillin.[2]

By 1947, the first cases of antibiotic resistance were reported. How did Dr. Fleming so confidently predict this development? He understood homeostasis. Exposure causes resistance. A biological system that becomes disturbed tries to go back to its original state. As we use an antibiotic more and more, organisms resistant to it are naturally selected to survive and reproduce. Eventually, these resistant organisms dominate, and the antibiotic becomes useless.

To prevent the development of antibiotic resistance, we must severely curtail the use of antibiotics. Unfortunately, the knee-jerk reaction of many doctors to antibiotic resistance is to use more antibiotics to 'overcome' the resistance—which backfires, since it only leads to more resistance. *Persistent, high-level use of antibiotics causes antibiotic resistance.*

VIRAL RESISTANCE

WHAT ABOUT VIRAL resistance? How do we become resistant to viruses like diphtheria, measles or polio for instance? Before the development of vaccines, it was viral infection itself that caused resistance to further infection. If you became infected with measles virus as a child, you'd be protected from reinfection with measles for the rest of your life. Most (though not all) viruses work this way. Exposure causes resistance.

Vaccines work on exactly this principle. Edward Jenner, working in rural England, heard the common tale of milkmaids developing resistance to the fatal smallpox virus because they had contracted the mild cowpox virus. In 1796, he deliberately infected a young boy with cowpox and observed how he was subsequently protected from smallpox, a similar virus. Through being inoculated with a dead or weakened virus, we build up immunity without actually causing the full disease. In other words, *viruses cause viral resistance.* Higher doses, usually in the form of repeated vaccinations, cause more resistance.

DRUG RESISTANCE

WHEN COCAINE IS taken for the first time, there is an intense reaction—the 'high.' With each subsequent use of the drug, the high becomes less intense. Sometimes users start to take larger and larger doses to achieve the same high. Through exposure to the drug, the body develops resistance to its effects—a condition called *tolerance.* People can build up tolerance to narcotics, marijuana, nicotine, caffeine, alcohol, benzodiazepines and nitroglycerin.

The mechanism of drug resistance is well known. To produce a desired effect, drugs, like hormones, are like keys that fit into the locks of the receptors on the cell surface. Morphine, for example, acts upon opioid receptors to provide pain relief. When there is a prolonged and excessive exposure to drugs, the body reacts by decreasing the number of receptors. Once again, the fundamental biological principle of homeostasis is at work here. If there is too much stimulation, the cell receptors are down-regulated, and the keys don't fit into the locks as well. The biological system returns closer to its original state. In other words, *drugs cause drug resistance.*

VICIOUS CYCLES

THE AUTOMATIC RESPONSE to the development of resistance is to increase the dosage. For example, in the case of antibiotic resistance, we respond by using more antibiotics. We use higher doses or newer drugs. The automatic response to drug resistance is to use more drugs. An alcoholic takes higher and higher doses of alcohol to beat the resistance, which temporarily 'overcomes' the resistance.

However, this behavior is clearly self-defeating. Since resistance develops in *response* to high, persistent levels, raising the dose in fact raises resistance. If a person uses larger amounts of cocaine, he or she develops greater resistance. As more antibiotics are used, more antibiotic resistance develops. This cycle continues until we simply can't go any higher.

And it's a self-reinforcing cycle—a vicious cycle. Exposure leads to resistance. Resistance leads to higher exposure. And the cycle keeps going around. Using higher doses has a paradoxical effect. The effect of using more antibiotics is to make antibiotics less effective. The effect of using more cocaine is to make cocaine less effective.

So let's recap what we know:

- Antibiotics cause antibiotic resistance. High doses cause more resistance.
- Viruses cause viral resistance. High doses cause more resistance.
- Drugs cause drug resistance (tolerance). High doses cause more resistance.

Now let's go back and ask our original question—what causes insulin resistance?

INSULIN CAUSES INSULIN RESISTANCE

IF INSULIN RESISTANCE is similar to other forms of resistance, the first thing to look at is *high, persistent levels of insulin itself.* If we increase insulin levels, do we get insulin resistance? That's an easy hypothesis to test—and luckily, studies have already been conducted on it.

SUPPORTING EVIDENCE

AN INSULINOMA IS a rare tumor[3,4] that secretes abnormally large amounts of insulin in the absence of any other significant disease. As the patient's insulin levels increase, his or her levels of insulin resistance increase in lock step—a protective mechanism and a very good thing. If insulin resistance did not develop, the high insulin levels would rapidly lead to very, very low blood sugars. The resulting severe hypoglycemia would quickly lead to seizures and death. Since the body doesn't want to die (and neither do we), it protects itself by developing insulin resistance—demonstrating homeostasis. The resistance develops naturally to shield against the unusually large insulin levels. *Insulin causes insulin resistance.*

Surgery to remove the insulinoma is the preferred treatment and dramatically lowers the patient's insulin levels. With the tumor gone, insulin resistance is also dramatically reversed, as well as associated conditions.[5] So reversing the high insulin levels reverses insulin resistance.

It is a simple matter to experimentally replicate the condition of an insulinoma. We can infuse higher-than-normal levels of insulin into a group of normal, healthy, non-diabetic volunteers. Can we induce insulin resistance?[6] Absolutely. A forty-hour insulin infusion reduced the subjects' ability to use glucose by a significant 15 per cent. Put another way, they developed 15 per cent greater insulin resistance. Here's the implication of this finding: I can make you insulin resistant. I can make *anybody* insulin resistant. All I need to do is give insulin.

Even using normal, physiologic levels of insulin will yield the exact same result.[7] Men with no previous history of obesity, pre-diabetes or diabetes were given a ninety-six-hour constant intravenous infusion of insulin. By the end, their insulin sensitivity dropped by 20 per cent to 40 per cent. The implications are simply staggering. With normal but persistent amounts of insulin alone, these healthy, young, lean men can be made insulin resistant. I can start these men on the road to diabetes and obesity simply by administering insulin—*which causes insulin resistance.* In the normal situation, of course, insulin levels do not remain persistently elevated like that.

Insulin is most often prescribed in type 2 diabetes to control blood sugars, sometimes in very high doses. Our question is, 'Do large doses of insulin cause insulin resistance?'

A 1993 study measured this effect.[8] Patients were started on intensive insulin treatment. In six months, they went from no insulin to 100 units a day on average. Their blood sugars were very, very well controlled. But the more insulin they took, the more insulin resistance they got—a direct causal relationship, as inseparable as a shadow is from a body. Even as their sugars got better, *their diabetes was getting worse!* These patients also gained an average of approximately 19 pounds (8.7 kilograms), *despite reducing their calorie intake by 300 calories per day.* It didn't matter. Not only does insulin cause insulin resistance, it also causes weight gain.

TIME DEPENDENCE AND OBESITY

SO WE KNOW that insulin causes insulin resistance. But insulin resistance also causes high insulin—a classic vicious or self-reinforcing, cycle. The higher the insulin levels, the greater the insulin resistance. The greater the resistance, the higher the levels. The cycle keeps going around and around, one element reinforcing the other, until insulin is driven up to extremes. *The longer the cycle continues, the worse it becomes*—that's why obesity is so time dependent.

People who are stuck in this vicious cycle for decades develop significant insulin resistance. That resistance leads to high insulin levels that are *independent of that person's diet.* Even if you were to change your diet, the resistance would still keep your insulin levels high. If your insulin levels stay high, then your body set weight stays high. The thermostat is set high, and your weight will be drawn irresistibly upward.

The fat get fatter. The longer you are obese, the harder it is to eradicate. But you already knew that. Oprah knew it. Everybody already knew it. Most current theories of obesity cannot explain this effect, so they instead ignore it. But obesity is *time-dependent.* Like rust, it takes time to develop. You can study moisture conditions and metal composition. But if you ignore the time-dependent nature of rust, you will not understand it.

A diet high in foods that provoke an insulin response may initiate obesity, but over time, insulin resistance becomes a larger and larger part of the problem and can become, in fact, a major driver of high insulin levels. Obesity drives itself. A long-standing obesity cycle is extremely difficult to break, and *dietary changes alone may not be sufficient.*

WHICH CAME FIRST?

THERE IS AN interesting chicken-and-egg problem here. High insulin leads to insulin resistance, and insulin resistance leads to high insulin. So which one comes first? High insulin or strong insulin resistance? Both are possible. But the answer can be found by following the time course of obesity.

In a 1994 study, researchers compared three groups of patients: non-obese, recently obese (less 4.5 years) and long-standing obese (more than 4.5 years).[9] The non-obese had lower insulin levels. This finding is expected. But both groups of obese subjects had equally high insulin levels, meaning that these levels go up but do not continue to go up over time.

What about insulin resistance? As the very beginning of obesity, a person will manifest little insulin resistance, but it *develops over time.* The longer you are obese, the more insulin resistance you have. Gradually, that insulin resistance will cause even your fasting insulin levels to rise.

The high insulin levels are the primary insult. Persistent high insulin levels lead gradually and eventually to insulin resistance. Insulin resistance in turn leads to higher insulin levels. But the crucial starting point of the vicious cycle is high insulin levels. Everything else follows and develops with time—and the fat get fatter.

COMPARTMENTALIZATION OF INSULIN RESISTANCE

HOW DOES INSULIN resistance produce obesity? We know that the hypothalamic area of the brain controls the body set weight and that insulin plays a key role in resetting the body set weight up or down. As insulin resistance develops, does it develop in all the cells in the body, including the brain? If all cells are insulin resistant, then high levels of it should not increase the body set weight. However, all the cells in the body are not equally resistant. Insulin resistance is compartmentalized.

The main compartments are the brain, liver and muscle. Changing the resistance of one does not change resistance in the others. For example, hepatic (liver) insulin resistance does not affect insulin resistance in the brain or muscle. When we ingest excess carbohydrates, we develop hepatic insulin resistance. Significant dietary intervention will reverse the hepatic insulin resistance, but will have no effect on insulin resistance in the muscles or the brain. Lack of exercise may lead to insulin resistance in the muscles. Exercise will increase insulin sensitivity there, but has little effect on insulin resistance in the liver or brain.

In response to hepatic or muscle insulin resistance, overall insulin levels increase. However, at the appetite centers in the hypothalamus, insulin's effect is unchanged. The brain is not resistant to insulin. When high insulin levels reach the brain, the insulin retains its full effect to raise body set weight.

PERSISTENCE CREATES RESISTANCE

HIGH HORMONAL LEVELS *by themselves* cannot cause resistance. Otherwise, all of us would quickly develop crippling resistance. We are naturally defended against resistance because we secrete our hormones—cortisol, insulin, growth hormone, parathyroid hormone or any other hormone—in bursts. High levels of hormones are released at specific times to produce a specific effect. Afterward, the levels quickly drop and stay very low.

Consider the body's daily rhythm. The hormone melatonin, produced by the pineal gland, is virtually undetectable during the day. As night falls, it begins to increase, and its levels peak in the early morning hours. Cortisol levels also rise in the early morning hours and spike just before we awaken. Growth hormone is secreted mostly in deep sleep and is usually undetectable during the day. Thyroid-stimulating hormone peaks in early morning. The periodic release of all these hormones is *essential* in preventing resistance.

Whenever the body is exposed to a constant stimulus, it acclimates to it (once again, homeostasis at work). Have you ever watched a baby sleep in a crowded, noisy airport? The ambient noise is very loud, but constant. The baby adapts by developing resistance to the noise. It basically just ignores it. Now imagine the same baby sleeping in a quiet house. A slight creak of the floorboards may be enough to wake him up. Even though it is not loud, it is very noticeable. The baby isn't used to the noise. High persistent levels create resistance.

Hormones work in exactly the same way. Most of the time, hormone levels are low. Every so often, a brief pulse of hormone (thyroid, parathyroid, growth, insulin—whatever) comes along. After it passes, levels are very low again. By cycling between low and high levels, the body never gets a chance to adapt. The brief pulse of hormone is over long before resistance develops.

What our body does, in effect, is to continually keep us in a quiet room. Every once in a while, we are momentarily exposed to a sound. Each time this happens, we experience the full effect. We are never given a chance to get accustomed to it—to develop resistance.

High levels alone do not lead to resistance. There are two requirements for resistance—high hormonal levels and constant stimulus. We've known this for quite some time. In fact, we use this to our advantage in drug therapy for angina (chest pain). Patients prescribed a nitroglycerin patch are often given the instructions to put the patch on in the morning and take it off in the evening.

By alternating periods of high drug effect and low drug effect, there is no chance for the body to develop resistance to the nitroglycerin. If the drug patch is worn constantly, it quickly becomes useless. Our body simply develops drug resistance.

How does this apply to insulin and obesity?

Consider the experiment described earlier that used constant infusions of insulin. Even healthy young men quickly developed insulin resistance. But the levels of insulin administered were normal. What changed? *The periodic release.* Normally, insulin is released in bursts, which prevents the development of insulin resistance. In the experimental condition, the constant bombardment of insulin led the body to down regulate its receptors and develop insulin resistance. Over time, insulin resistance induces the body to produce even more insulin to 'overcome' the resistance.

In the case of insulin resistance, it comes down to both meal composition and meal timing—the two critical components of insulin resistance. The types of food eaten influence the insulin levels. Should we eat candy or olive oil? This is the question of macronutrient composition, or 'what to eat.' However, the persistence of insulin plays a key role in the development of insulin resistance, so there is also the question of meal timing, or 'when to eat.' Both components are equally important. Unfortunately, we spend obsessive amounts of time and energy trying to understand what we should be eating and devote virtually no time to when we should be eating. We are only seeing half the picture.

THREE MEALS A DAY. NO SNACKS.

LET'S TURN BACK the clock to the U.S. in the 1960s. Food shortages from the war are a thing of the past. Obesity is not yet a major issue. Why not? After all, they ate Oreo cookies, KitKats, white bread and pasta. They ate sugar, although not quite as much. They also ate three meals per day, with no snacks in between.

Let's assume breakfast is taken at 8 a.m. and dinner at 6 p.m. That means that they have balanced ten hours of eating with fourteen hours of fasting. The periods of increased insulin (feeding) are balanced by periods of decreased insulin (fasting).

Eating large amounts of refined carbohydrates like sugar and white bread makes for higher insulin peaks. So why was obesity slow to progress? The decisive difference is that there was a daily period of low insulin levels. Insulin resistance requires *persistently* high levels. The nightly fasting caused periods of very low insulin, so *resistance could not develop.* One of the key factors in obesity's development was removed.

Figure 10.1. Insulin release with an eating pattern of three meals, no snacks.

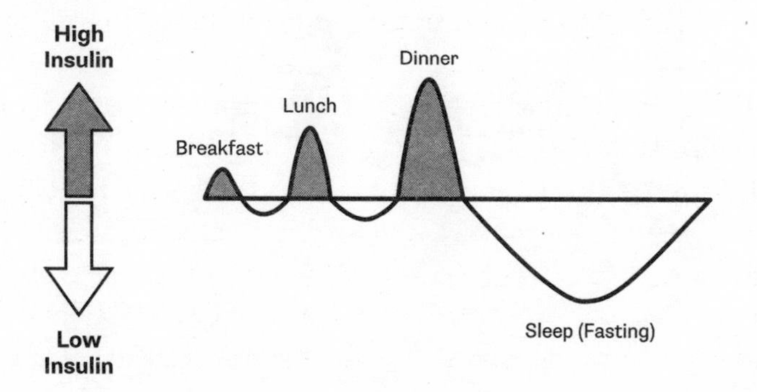

Pulses of insulin (mealtimes) are followed by a long fasting period (sleep), as illustrated in Figure 10.1. However, the situation changes entirely when we are *constantly* exposed to insulin. What would happen if daily eating opportunities are increased from three to six—which is exactly what's happened since the 1970s. Moms everywhere knew that eating snacks all the time was a bad idea: 'It'll make you fat'; 'You'll ruin your dinner.' But nutritional authorities have now decided that snacking is actually *good* for us. That eating *more* often will make us thinner, as ridiculous as that sounds. Many obesity specialists and physicians suggest eating even more frequently, every 2.5 hours.

An American survey of more than 60,000 adults and children revealed that, in 1977, most people ate three times a day.[10] By 2003, most people were eating five to six times a day. That is, three meals a day plus two to three snacks in between. The average time between meals has dropped 30 per cent, from 271 minutes to 208 minutes. The balance between the fed state (insulin dominant) and the fasted state (insulin deficient) has been completely destroyed. (See Figure 10.2.) We now spend most of our time in the fed state. Is it any great mystery that we're gaining weight?

Figure 10.2. Insulin release with an eating pattern of multiple meals and snacks.

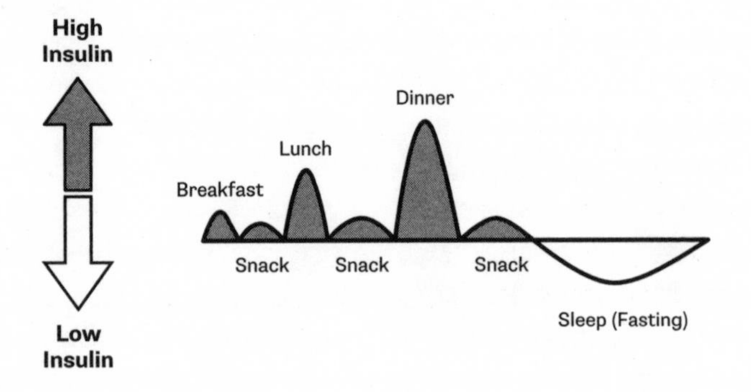

But the story gets worse. Insulin resistance, in turn, leads to higher *fasting* insulin levels. Fasting insulin levels are normally low. Now, instead of starting the day with low insulin after the nightly fast, we are starting with high insulin. The persistence of high insulin levels leads to even more resistance. In other words, insulin resistance itself leads to more resistance—a vicious cycle.

We have now fulfilled the two prerequisites of insulin resistance—high levels and persistence. Following a low-fat diet led to the inadvertent increase in refined-carbohydrate consumption, which stimulates high levels of insulin, which contributes to weight gain.

But in the development of obesity, the increase in meals is almost twice as important as the change in diet.[11] We obsess about what we should eat. We eat foods that practically didn't exist ten years ago. Quinoa. Chia seeds. Acai berries. All in the hopes of making us slim. But we spare not even a single thought as to *when* we should be eating.

Several myths are often perpetuated to convince people that snacking is beneficial. The first myth is that eating frequently will increase your metabolic rate. Your metabolic rate does increase slightly after meals to digest your food—the thermogenic effect of food. However, the overall difference is extremely small.[12] Eating six small meals per day causes the metabolic rate to go up six times a day, but only a little. Eating three larger meals per day causes metabolic rate to go up three times a day, but a lot each time. In the end, it's a wash. The total thermogenic effect of food over twenty-four hours for both the grazing and gorging strategies is the same: neither yields a metabolic advantage. Eating more frequent meals does not aid in weight loss.[13]

The second myth is that eating frequently controls hunger, but evidence is impossible to find. Once people decided that grazing was better, I suppose all sorts of reasons were invented to justify it. Recent studies don't support this notion.[14]

The third myth is that eating frequently keeps blood glucose from becoming too low. But unless you have diabetes, your blood sugars are stable whether you eat six times a day or six times a month. People have fasted for prolonged periods without low blood sugar, the world record being 382 days.[15] The human body has evolved mechanisms to deal with prolonged periods without food. The body instead burns fat for energy, and blood sugar levels remain in the normal range, even during prolonged fasting, due to gluconeogenesis.

We are eating all the time. Societal norms, which had previously frowned upon eating except at mealtimes, now permit eating anywhere, anytime. Government agencies and schools actively encourage snacking, something that previously had been heavily discouraged. We are taught to eat the minute we roll out of bed. We are taught to eat throughout the day and eat again just before sleep. We spend up to eighteen hours in the insulin-dominant state, with only six hours insulin deficient. Figure 10.3, illustrates how much the balance between the insulin-dominant and insulin-deficient states has changed.

Figure 10.3. The balance of time spent each day in the insulin-dominant versus the insulin-deficient state has changed greatly since the 1970s.

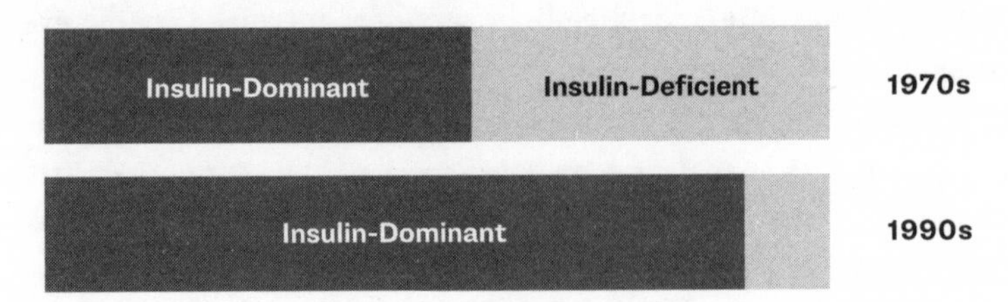

Crazier still—we have been brainwashed to believe that constant eating is somehow *good* for us! Not just acceptable, but *healthy.*

In order to accommodate all those eating opportunities, societal norms have also changed. Previously, all eating was done at mealtimes at a table. Now, it is acceptable to eat anywhere. We can eat in the car. We can eat in the movie theatre. We can eat in front of the TV. We can eat in front of the computer. We can eat while walking. We can eat while talking. We can eat in a box. We can eat with a fox. We can eat in a house. We can eat with a mouse. You get the picture.

Millions of dollars are spent to give children snacks all day long. Then millions more are spent to combat childhood obesity. These same kids are berated for getting fat. Millions more are spent to fight obesity as adults.

The increase in eating opportunities has led to persistence of high levels of insulin. Snacks, which tend to be high in refined carbohydrates, also tend to cause high levels of insulin. Under these conditions, we should *expect* the development of insulin resistance.

We never consider the implications of the drastic changes we have made in meal timing. Think about it this way: In 1960, we ate three meals a day. There wasn't much obesity. In 2014, we eat six meals a day. There is an obesity epidemic.

So, do you really think we should eat six meals day? While movies such as *Super Size Me* get all the headlines, and while people screech about portion control, the main culprit lies completely hidden—the insidious snack. Indeed, many health professionals have been very vocal about *increasing* the number of eating occasions. This situation is just as crazy as it sounds. Eat more to weigh less. That doesn't even sound like it will work.

And guess what? It doesn't.

PART FOUR

The Social Phenomenon of Obesity

(11)

BIG FOOD, MORE FOOD AND THE NEW SCIENCE OF DIABESITY

•

FUELING THE INCREASE in eating opportunities was the desire of big food companies to make more money. They created an entirely new category of food, called 'snack food,' and promoted it relentlessly. They advertised on TV, print, radio and Internet.

But there is an even more insidious form of advertising called sponsorship and research. Big Food sponsors many large nutritional organizations. And then there are the medical associations. In 1988, the American Heart Association decided that it would be a good idea to start accepting cash to put its Heart Check symbol on foods of otherwise dubious nutritional quality. The Center for Science in the Public Interest estimates that in 2002, the AHA received over $2 million from this program alone.[1] Food companies paid $7500 for one to nine products, but there was a volume discount for more than twenty-five products! Exclusive deals were, of course, more expensive. In 2009, nutritional standouts such as Cocoa Puffs and Frosted Mini Wheats were still on the Heart Check list. The 2013 Dallas Heart Walk organized by the AHA featured Frito-Lay as a prominent sponsor. The Heart and Stroke Foundation in Canada was no better. As noted on Dr. Yoni Freedhoff's blog, a bottle of grape juice proudly bearing the Health Check contained ten teaspoons of sugar.[2] The fact that these foods were pure sugar seemed not to bother anybody.

Researchers and academic physicians, as key opinion leaders, were not to be ignored either. Many health professionals endorse the use of artificial meal-replacement shakes or bars, drugs and surgery as evidence-based diet aids. Forget about eating a whole, unrefined natural-foods diet. Forget about reducing added sugars and refined starches such as white bread. Consider the ingredient list of a popular meal-replacement shake. The first five ingredients are water, corn maltodextrin, sugar, milk protein concentrate and canola oil. This nauseating blend of water, sugar and canola oil does not really meet my definition of healthy.

In addition, impartiality—or the lack thereof—can be a serious issue when it comes to publishing medical and health information. The financial-disclosures section of some papers published in journals and on the web can run for more than half a page. Funding sources have enormous influence on study results.[3] In a 2007 study that looked specifically at soft drinks, Dr. David Ludwig from Harvard University found that accepting funds from companies whose products are reviewed increased the likelihood of a favorable result by approximately 700 per cent! This finding is echoed in the work of Marion Nestle, professor of nutrition and food studies at New York University. In 2001, she concluded that it is 'difficult to find studies that did not come to conclusions favoring the sponsor's commercial interest.'[4]

The fox, it seemed, was now guarding the hen house. Shills for Big Food had been allowed to infiltrate the hallowed halls of medicine. Push fructose? No problem. Push obesity drugs? No problem. Push artificial meal replacement shakes? No problem.

But the obesity epidemic couldn't very well be ignored, and a culprit *had* to be found. 'Calories' was the perfect scapegoat. Eat fewer calories, they said. But eat more of everything else. There is no company that sells 'Calories,' nor is there a brand called 'Calories.' There is no food called 'Calories.' Nameless and faceless, calories were the ideal stooge. 'Calories' could now take all the blame.

They say candy doesn't make you fat. Calories make you fat. They say that 100 calories of cola is just as likely as 100 calories of broccoli to make you fat. They say that a calorie is a calorie. Don't you know? But show me a single person that grew fat by eating too much steamed broccoli. I know it. You know it.

Furthermore, we cannot simply eat our usual diet and add some fat or protein or snacks and expect to lose weight. Against all common sense, weight-loss advice usually involves *eating more.* Just take a look at Table 11.1.

Table 11.1. Conventional advice for weight loss.

Eat 6 times a day
Eat high protein
Eat more vegetables
Eat more omega 3s
Eat more fiber
Eat more vitamins
Eat more snacks
Eat low fat
Eat breakfast
Eat more calcium
Eat more whole grains
Eat more fish

Why would anybody give such completely asinine advice? *Because nobody makes any money when you eat less.* If you take more supplements, the supplement companies make money. If you drink more milk, the dairy farmers make money. If you eat more breakfast, the breakfast-food companies make money. If you eat more snacks, the snack companies make money. The list goes on and on. One of the worst myths is that eating more frequently causes weight loss. Eat snacks to lose weight? It *sounds* pretty stupid. And it is.

SNACKING: IT WON'T MAKE YOU THIN

HEALTH PROFESSIONALS NOW heavily promote snacking, which previously had been heavily discouraged. But studies confirm that snacking means you eat more. Subjects given mandatory snacks[5] would consume slightly fewer calories at the subsequent meal, but not enough to offset the extra calories of the snack itself. This finding held true for both fatty and sugary snacks. Increasing meal frequency does not result in weight loss.[6] Your grandmother was right. Snacking will make you fat.

Diet quality also suffers substantially because snacks tend to be very highly processed. This fact mainly benefits Big Food, since selling processed instead of real foods yields a much larger profit. The need for convenience and shelf life lends itself to refined carbohydrates. After all, cookies and crackers are mostly sugar and flour—and they don't spoil.

BREAKFAST: THE MOST IMPORTANT MEAL TO SKIP?

THE MAJORITY OF Americans identify breakfast as the most important meal of the day. Eating a hearty breakfast is considered a cornerstone of a healthy diet. Skipping it, we are told, will make us ravenously hungry and prone to overeat for the rest of the day. Although we think it's a universal truth, it's really only a North American custom. Many people in France (a famously skinny nation) drink coffee in the morning and skip breakfast. The French term for breakfast, petit déjeuner (little lunch) implicitly acknowledges that this meal should be kept small.

The National Weight Control Registry was established in 1994 and monitors people who have maintained a weight loss of 30 pounds (14 kilograms) for more than one year. The majority (78 per cent) of the National Weight Control Registry participants eat breakfast.[7] This, we are told, is proof that eating breakfast aids weight loss. But what percentage of those who did *not* lose weight ate breakfast? Without knowing, it's impossible to draw any firm conclusions. What if 78 per cent of those that did not lose weight also ate breakfast? This data is not available.

Furthermore, the National Weight Control Registry itself is a highly self-selected population and not representative of the general population.[8] For example, 77 per cent of registrants are women, 82 per cent are college educated and 95 per cent are Caucasian. Furthermore, an association (for instance, between weight loss and eating breakfast) does not mean causality. A 2013 systematic review of breakfast eating found that most studies interpreted the available evidence in favor of their own bias.[9] Authors who previously believed that breakfast protected against obesity interpreted the evidence as supportive. In fact, there are few controlled trials, and most of those show no protective effect from eating breakfast.

It is simply not necessary to eat the minute we wake up. We imagine the need to 'fuel up' for the day ahead. However, our body has already done that automatically. Every morning, just before we wake up, a natural circadian rhythm jolts our bodies with a heady mix of growth hormone, cortisol, epinephrine and norepinephrine (adrenalin). This cocktail stimulates the liver to make new glucose, essentially giving us a shot of the good stuff to wake us up. This effect is called the dawn phenomenon, and it has been well described for decades.

Many people are not hungry in the morning. The natural cortisol and adrenalin released stimulates a mild flight-or-fight response, which activates the sympathetic nervous system. Our bodies are gearing up for action in the morning, not for eating. All these hormones release glucose into the blood for quick energy. We're already gassed up and ready to go. There is simply no need to refuel with sugary cereals and bagels. Morning hunger is often a behavior learned over decades, starting in childhood.

The word breakfast literally means the meal that *breaks* our *fast*, which is the period when we are sleeping and therefore not eating. If we eat our first meal at 12 noon, then grilled salmon salad will be our 'break fast' meal—and there's nothing wrong with that.

A large breakfast is thought to reduce food intake throughout the rest of the day. However, such does not always seem to be the case.[10] Studies show that lunch and dinner portions tend to stay constant, regardless of the amount of calories taken at breakfast. The more one eats at breakfast, the higher the total caloric intake over the entire day. Worse, taking breakfast increases the number of eating opportunities in a day. Breakfast eaters therefore tend to eat more and eat more often—a deadly combination.[11]

Furthermore, many people confess that they are not hungry first thing in the morning and force themselves to eat only because they feel that doing so is the healthy choice. As ridiculous as it sounds, many people force themselves to eat more in an effort to lose weight. In 2014, a sixteen-week randomized controlled trial of breakfast eating found that 'contrary to widely espoused views this had no discernable effect on weight loss.'[12]

We are often told that skipping breakfast will shut down our metabolism. The Bath Breakfast Project, a randomized controlled trial, found that 'contrary to popular belief, there was no metabolic adaptation to breakfast.'[13] Total energy expenditure was the same whether one ate breakfast or not. Breakfast eaters averaged 539 extra calories per day compared to those that skipped breakfast—a finding consistent with other trials.

The main problem in the morning is that we are always in a rush. Therefore, we want the convenience, affordability and shelf life of processed foods. Sugary cereals are the kings of the breakfast table, with children as the primary target. The vast majority (73 per cent) of children regularly eat sugary cereals. By contrast, only 12 per cent regularly eat eggs at breakfast. Other easy-to-prepare foods like toast, bread, sugary yogurts, Danishes, pancakes, donuts, muffins, instant oatmeal and fruit juice are also popular. Clearly, the cheap refined carbohydrate reigns supreme here.

Breakfast is the most important meal of the day—for Big Food. Sensing the perfect opportunity to sell more highly profitable, highly processed 'breakfast' foods, Big Food circled the easy money like sharks on wounded prey. 'Eat breakfast!' they thundered. 'It's the most important meal of the day!' they bellowed. Even better, here was an opportunity to 'educate' the doctors, dieticians and other medical professionals. Those people had a respectability Big Food could never achieve. So the money flowed.

There are some commonsense questions you can ask yourself about breakfast. Are you hungry at breakfast? If not, listen to your body and don't eat. Does breakfast *make* you hungry? If you eat a slice of toast and drink a glass of orange juice in the morning—are you hungry an hour later? If so, then don't eat breakfast. If you *are* hungry and want to eat breakfast, then do so. But avoid sugars and refined carbohydrates. Skipping breakfast does not give you the freedom to eat a Krispy Kreme donut as a mid-morning snack either.

FRUITS AND VEGETABLES: THE FACTS

ONE OF THE most pervasive pieces of weight-loss advice is to eat more fruits and vegetables, which are undeniably relatively healthy foods. However, if your goal is to lose weight, then it logically follows that deliberately eating more of a healthy food is not beneficial unless it replaces something else in your diet that is less healthy. However, nutritional guidelines don't state this. For example, the World Health Organization writes, 'Prevention of obesity implies the need to: Promote the intake of fruits and vegetables.'[14]

The 2010 *Dietary Guidelines for Americans* also stresses the importance of increasing consumption of fruits and vegetables. In fact, this recommendation has been part of *Dietary Guidelines* since its very inception. Fruits and vegetables are high in micronutrients, vitamins, water and fiber. They may also contain antioxidants and other healthful phytochemicals. What is not explicit is that increased intake should displace less healthy foods in our diet. It's assumed that with the low-energy density and high fiber of fruits and vegetables, our satiety will increase, and therefore we'll eat less of calorie-rich foods. If this strategy is the main mechanism of weight loss, our advice should be to 'replace bread with vegetables.' But it is not. Our advice is simply to eat more fruits and vegetables. Can we really eat more to lose weight?

In 2014, researchers gathered all available studies on increased intake of fruit-and-vegetable and weight loss.[15] They could not find a single study to support this hypothesis. Combining all the studies did not show any weight-loss benefit either. To put it simply, you cannot eat more to weigh less, even if the food you're eating more of is as healthy as vegetables.

So should we eat more fruits and vegetables? Yes, definitely. *But only if they are replacing other unhealthier foods in your diet.* Replace. Not add.[16]

THE NEW SCIENCE OF DIABESITY

EXCESSIVELY HIGH INSULIN resistance is the disease known as type 2 diabetes. High insulin resistance leads to elevated blood sugars, which are a symptom of this disease. In practical terms, this means that not only does *insulin causes obesity,* but also that *insulin causes type 2 diabetes.* The common root cause of both diseases is high, persistent insulin levels. Both are diseases of hyper-insulinemia (high insulin levels). Because they are so similar, both diseases are beginning to be observed as a syndrome, aptly termed diabesity.

That high insulin levels cause both obesity and type 2 diabetes has profound implications. The treatment for *both* is to *lower* insulin levels, yet current treatments focus on *increasing* insulin levels, which is exactly wrong. Giving insulin for type 2 diabetes will worsen, not improve, the disease. But can lowering insulin levels cure type 2 diabetes? Absolutely. But the many misunderstandings about type 2 diabetes would require another book to clarify.

Our own disastrous, misguided dietary changes since the 1970s have created the diabesity debacle. We have seen the enemy, and it is ourselves. Eat more carbohydrates. Eat more often. Eat breakfast. Eat more. Ironically, these dietary changes were prescribed to reduce heart disease, but instead, we've encouraged it since diabesity is one of the strongest risk factors for heart disease and stroke. We've been trying to put out a fire with gasoline.

(12)

POVERTY AND OBESITY

•

THE CENTER FOR Disease Control in Atlanta keeps detailed statistics about the prevalence of obesity in the United States, which varies strikingly between regions. It is also quite notable that those states with the *least* obesity in 2010 nonetheless have higher rates than those that were found in states with the *most* obesity in 1990. (See Figure 12.1.[1])

Overall, there has been a huge increase in obesity in the United States. Despite the similarity in culture and genetics between populations in Canada and the United States, U.S rates of obesity are much higher. This fact suggests that government policies must play a role in the development of obesity. Southern states such as Texas tend to have much more obesity than those in the west (California, Colorado) and north east.

Socioeconomic status has long been known to play a role in the development of obesity in that poverty correlates very closely with obesity. States with the most poverty tend to also have the most obesity. The southern states are relatively less affluent than those in the west and north east. With a 2013 median income of $39,031, Mississippi is the poorest state in the U.S.[2] It also has the highest level of obesity, at 35.4 per cent.[3] But why is poverty linked to obesity?

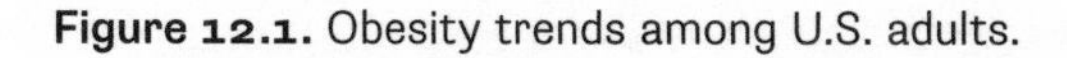
Figure 12.1. Obesity trends among U.S. adults.

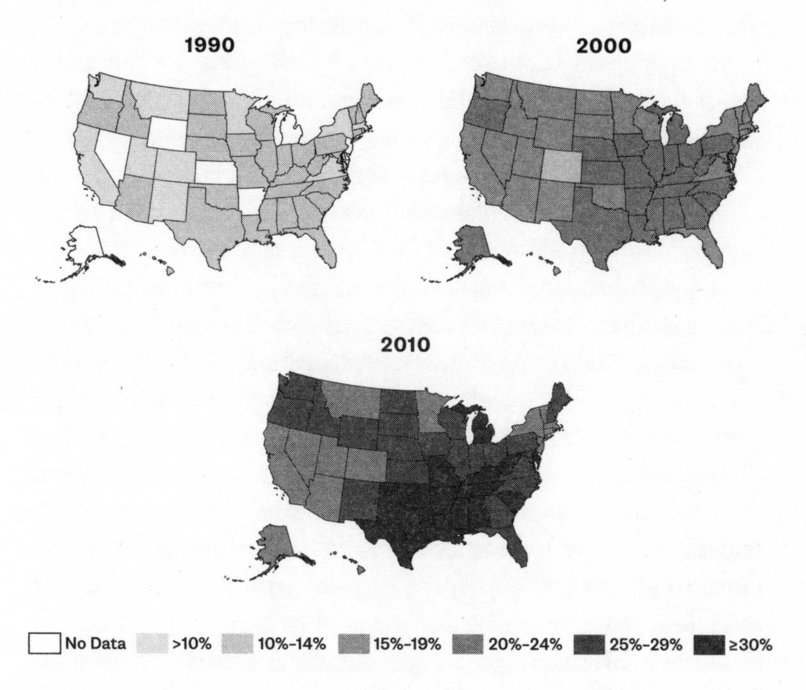

THEORIES, CALORIES AND THE PRICE OF BREAD

THERE IS A theory of obesity called the food-reward hypothesis, which postulates that the rewarding quality of the food causes overeating. Maybe obesity rates have increased because food is more enjoyable than it has ever been, causing people to eat more. Rewards reinforce behavior, and the behavior of eating is rewarded by the palatability—the deliciousness—of the food.

The increased palatability of food is not accidental. Societal changes have resulted in more meals being eaten away from the home, at restaurants and fast-food outlets. Many foods prepared in those venues may be specifically engineered to be hyper-palatable through the use of chemicals, additives and other artificial processes. The addition of sugar and seasonings such as monosodium glutamate (MSG) may trick the taste buds into believing that the food is more rewarding.

This argument is put forth in books such as *Sugar, Salt and Fat: How the Food Giants Hooked Us*,by Michael Moss[4], and *The End of Overeating: Taking Control of the Insatiable American Appetite*, by David Kessler.[5] Added sugars, salt and fat and their combination bear a disproportionate amount of blame for inducing us to overeat. But people have been eating salt, sugar and fat for the last 5000 years. These are not new additions to the human diet. Ice cream, a combination of sugar and fat, has been a summertime treat for more than 100 years. Chocolate bars, cookies, cakes and sweets existed long before the obesity epidemic in the 1970s. Children were enjoying their Oreo cookies in the 1950s without the problem of obesity.

The basic premise of this argument is that food is more delicious in 2010 than in 1970 because food scientists engineer it to be so. We cannot help but overeat calories and therefore become obese. The implication is that hyper-palatable 'fake' foods are more delicious and more rewarding than real foods, but that seems very difficult to believe. Is a 'fake,' highly processed food such as a TV dinner more delicious than fresh salmon sashimi dipped in soy sauce with wasabi? Or is Kraft Dinner, with its fake cheese sauce, really more enticing than a grilled rib-eye steak from a grass-fed cow?

But the association of obesity with poverty presents a problem. The food-reward hypothesis would predict that obesity should be more prevalent among the rich, since they can afford to buy more of the highly rewarding foods. But the exact opposite is true. Lower-income groups suffer more obesity. To be blunt, the rich can afford to buy food that is both rewarding and expensive, whereas the poor can afford only rewarding food that is cheaper. Steak and lobster are highly rewarding—and expensive—foods. Restaurant meals, which are expensive compared to home cooking, are also highly rewarding. Increased prosperity results in increased access to different types of highly rewarding food, which should result in more obesity. But it does not.

If this situation is not the result of diet, then perhaps the problem is lack of exercise. Perhaps the rich can afford to join gyms and therefore are more physically active, experiencing less obesity. In a similar vein, perhaps affluent children are more able to participate in organized sports, leading to less obesity. While these ideas may sound reasonable at first, further reflection reveals many discrepancies. The majority of exercise is free, often requiring no more than a basic shoe. Walking, running, soccer, basketball, push-ups, sit-ups and calisthenics all require minimal or no cost, and they are all excellent forms of exercise. Many occupations, such as construction or farming, involve significant physical exertion throughout the working day. Those jobs require heavy lifting, day after day after day. Contrast that to an office-bound lawyer or a Wall Street investment banker. Spending up to twelve hours a day perched in front of a computer, his or her physical exertion is limited to walking from desk to elevator. Despite this large difference in daily physical activity, obesity rates are higher in the less affluent but more physically active group.

Neither food reward nor physical exertion can explain the association between obesity and poverty. So what drives obesity in the poor? It is the same thing that drives obesity everywhere else: *refined carbohydrates.*

For those dealing with poverty, food needs to be affordable. Some dietary fats are fairly inexpensive. However, we do not, as a general rule, drink a cup of vegetable oil for dinner. Furthermore, official government recommendations are to follow a low-fat diet. Dietary proteins, such as meat and dairy, tend to be relatively expensive. Less expensive vegetable proteins, such as tofu or legumes, are available but not typical in a North American diet.

This leaves carbohydrates. If refined carbohydrates are significantly cheaper than other sources of food, then those living in poverty will eat refined carbohydrates. Indeed, processed carbohydrates are entire orders of magnitude less expensive. An entire loaf of bread might cost $1.99. An entire package of pasta might cost $0.99. Compare that to cheese or steak, which might cost $10 or $20. Unrefined carbohydrates, such as fresh fruits and vegetables, cannot compare to the low, low prices of processed foods. A single pound of cherries, for instance, may cost $6.99.

Why are highly refined carbohydrates so cheap? Why are unprocessed carbohydrates so much more expensive? The government lowers the cost of production with hefty agricultural subsidies. But not all foods get equal treatment. Figure 12.2 indicates which foods (and programs) receive the most in subsidies.[6]

Figure 12.2. U.S. agricultural subsidies, 1995-2012.

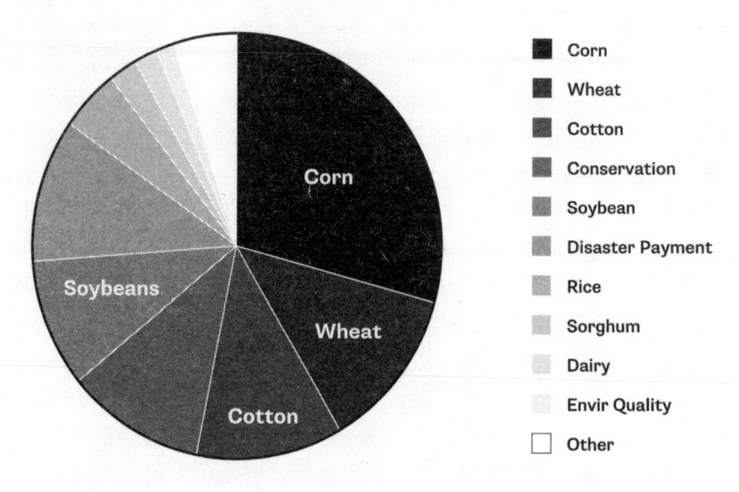

In 2011, the United States Public Interest Research Groups noted that 'corn receives an astounding 29 per cent of all U.S. agricultural subsidies, and wheat receives a further 12 per cent.'[7] Corn is processed into highly refined carbohydrates for consumption, including corn syrup, high-fructose corn syrup and cornstarch. Wheat is almost never consumed as a whole berry but further processed into flour and consumed in a wide variety of foods.

Unprocessed carbohydrates, on the other hand, receive virtually no financial aid. While mass production of corn and wheat receives generous support, the same cannot be said for cabbage, broccoli, apples, strawberries, spinach, lettuce and blueberries. Figure 12.3 compares the subsidy received for apples to that received for food additives, which include corn syrup, high-fructose corn syrup, corn starch and soy oils.[8] *Food additives receive almost thirty times more in subsidies.* Saddest of all, apples receive the *most,* not the *least,* federal aid of all the fruits and vegetables. All others receive negligible support.

Figure 12.3. Food additives are subsidized far more heavily than whole foods.

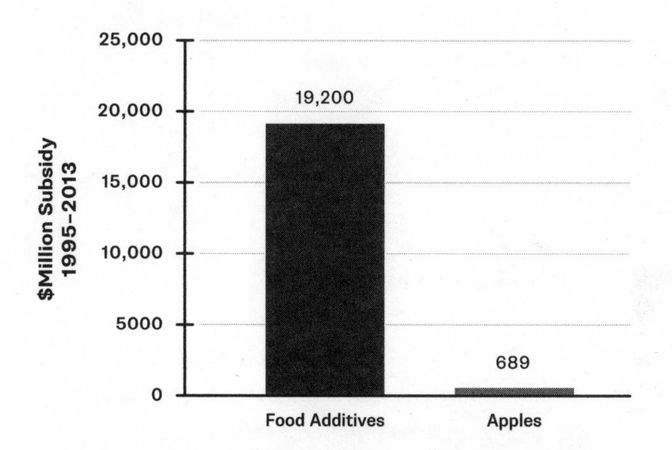

The government is subsidizing, with our own tax dollars, the very foods that are making us obese. *Obesity is effectively the result of government policy.* Federal subsidies encourage the cultivation of large amounts of corn and wheat, which are processed into many foods. These foods, in turn, become far more affordable, which encourages their consumption. Large-scale consumption of highly processed carbohydrates leads to obesity. More tax dollars are then needed to support anti-obesity programs. Even more dollars are needed for medical treatment of obesity-related problems.

Was this a giant conspiracy to keep us sick? Doubtful. The large subsidies were simply the result of programs to make food affordable, which began in earnest in the 1970s. Back then, the major health concern was not obesity, but the 'epidemic' of heart disease that was believed to be the result of excess dietary fat. The base of the Food Pyramid, the foods to be eaten by each of us every day, was bread, pasta, potatoes and rice. Naturally, money flowed into subsidies for those foods, the production of which was encouraged by the U.S. Department of Agriculture. Refined grains and corn products soon became affordable by all. Obesity followed like the grim reaper.

It is noteworthy that, in the 1920s, sugar was relatively expensive. A 1930 study showed that type 2 diabetes was far more common among the wealthier northern states compared to the poorer southern states.[9] As sugar became extremely cheap, however, this relationship inverted. Now, *poverty* is associated with type 2 diabetes, rather than the other way around.

EVIDENCE FROM THE PIMA PEOPLE

THE PIMA INDIANS of the American southwest have the highest rates of diabetes and obesity in North America. An estimated 50 per cent of Pima adults are obese, and of those, 95 per cent have diabetes.[10] High levels of obesity are once again seen alongside grinding poverty. What happened?

The traditional Pima diet relied on agriculture, hunting and fishing. All reports from the 1800s suggest that the Pima were 'sprightly' and in good health. By the early 1900s, American trading posts began to establish themselves. The Pimas' ancestral way of life, with its agriculture and hunting patterns, as well as its diet, was completely disrupted.

Refined carbohydrates, particularly white sugar and flour, began to replace traditional foods, since both substances could be stored at room temperature for long periods without spoilage. By the 1950s, obesity was widespread among the Pima in association with grinding poverty.

This situation is not unique to the Pima Indians. Obesity and diabetes have become a colossal health problem for virtually all native North American peoples, and the trend was already noticeable in the 1920s, decades before the current epidemic, which started in 1977.

Why? In times of plentiful natural whole foods, such as vegetables, wild game and fish, the Pima did not develop either obesity or diabetes. It was not until their traditional lifestyle and diet were disturbed that obesity became rampant.

It could be suggested that obesity is the result of a modern lifestyle, including the increased use of not just cars, but also computers, video games and laborsaving devices: the increasingly sedentary nature of our lifestyles could be the underlying cause of obesity.

On closer examination, this explanation holds water like a straw basket. Native American tribes developed obesity in the 1920s, decades before widespread use of cars. The North American obesity epidemic surged around 1977. But there was no corresponding surge in vehicle miles driven in 1977. There is only a steady increase from 1946 to 2007.[11, 12]

Other people suggest that increased prevalence of fast food may contribute to the obesity crisis. Again, there is no corresponding sharp upward spike in the number of restaurants, fast food or otherwise, in 1977. There is just a gradual increase over the decades. Similarly, obesity became rampant among the Pima decades before fast food became widespread. The surprise lies, in fact, that obesity became widespread among all native North American populations as early as the 1920s, even as the rest of North America was still relatively lean.

What explains the experience of the Pima? Simple enough. The same thing drives obesity in the Pima Indians as in everyone else: *highly refined carbohydrates.* When the Pima replaced traditional, unrefined foods with highly refined sugar and flour, they became obese. In 1977, the new *Dietary Guidelines* caused a sharp increase in the percentage intake of dietary carbohydrates. Obesity soon tagged along like a bratty little brother.

The hormonal obesity theory helps explain many apparent inconsistencies in the epidemiology of obesity. The driving factor in obesity is insulin, and in many cases, the wide availability of refined carbohydrates. This understanding helps explain an equally pressing issue: childhood obesity.

(13)

CHILDHOOD OBESITY

•

ALARMED BY THE stunning rise of obesity and type 2 diabetes in school-aged children, hundreds of millions of dollars were deployed to counterattack. The first choice in our arsenal was the beloved Eat Less, Move More approach, which sported a perfect record unblemished by success. Nevertheless, as nutritional authorities scrambled to do battle, only one diet plan got the call. The U.S. National Institutes of Health funded the HEALTHY study, a large three-year effort involving forty-two schools in grades six to eight.[1] Half of the schools would receive a multicomponent intervention, while the other half continued their usual routines. The study encouraged certain nutritional and exercise goals, including

- lowering average fat content of food,
- providing at least two servings of fruit and vegetables per student,
- providing at least two servings of grain-based food and/or legumes,
- limiting dessert and snack foods to less than 200 calories per item,
- limiting beverages to water, low-fat milk and 100 per cent fruit juice and
- encouraging more than 225 minutes of moderate to vigorous physical activity per week.

Our old friend—Eat Less, Move More. Not too bright, but as familiar as an old blanket. There were classroom-based programs, newsletters for parents, social marketing (branding, posters, in-school announcements), student events and incentives (T-shirts, water bottles). Both groups began with roughly 50 per cent of the students considered overweight or obese. By the end of three years, the Eat Less Move More group brought that down to 45 per cent. Success! The group that followed their usual habits finished at... 45 per cent. There was no measurable benefit for the diet and exercise group. *This weight-loss strategy was virtually useless.*

But who *hasn't* tried the Eat Less, Move More approach and failed? The HEALTHY study was only the latest in an unbroken string of failures.

OBESITY: NO LONGER JUST FOR ADULTS

DURING THE YEARS 1977 to 2000, the prevalence of childhood obesity skyrocketed in every age category. Obesity in children aged six to eleven increased from 7 per cent to 15.3 per cent. For children aged twelve to nineteen, it more than tripled, from 5 per cent to 15.5 per cent. Obesity-related diseases such as type 2 diabetes and high blood pressure, previously rare in children, are becoming more common. Obesity has metastasized from being solely an adult concern to being a pediatric one too.

Childhood obesity also leads to adult obesity and future health problems, particularly cardiovascular issues.[2] The Bogalusa Heart Study concluded, 'childhood obesity tracked into young adulthood,'[3] which is obvious to almost everybody. Childhood obesity is a predictor of increased mortality,but is, most importantly, a *reversible* risk factor.[4] Overweight children who became normal weight as adults have the same mortality risk as those who have never been overweight.[5]

Obesity has begun to afflict younger and younger children. In one study covering a twenty-two-year period ending in 2001, children of all ages show an increased prevalence of obesity, even in the zero- to six-month-old age range.[6]

That finding is especially interesting. Conventional calorie-based theories of obesity are unable to explain this trend. Obesity is considered an energy-balance problem, one of eating too much or exercising too little. Since six-month-olds eat on demand and are often breastfed, it is impossible that they eat too much. Since six-month-olds do not walk, it is impossible that they exercise too little. Similarly, birth weight has increased by as much as half a pound (200 grams) over the last twenty-five years.[7] The newborn cannot eat too much or exercise too little.

What is going on here?

Numerous hypotheses have been offered to explain newborn obesity. One popular theory suggests that certain chemicals (obesogens) in our modern environment lead to obesity, chemicals that are often endocrine disruptors. (That is, they disrupt the normal functional hormonal systems of the body.) Since obesity is a hormonal rather than a caloric imbalance, this notion does make some intuitive sense. Nonetheless, the majority of the data comes from animal studies.

For example, the pesticides atrazine and DDE may cause obesity in rodents.[8] However, no data is available for humans. Without such data, it is difficult to conclusively determine whether a chemical is an obesogen or not. Furthermore, studies use concentrations of chemicals that are hundreds or even thousands of times greater than normal human exposure. While these chemicals are almost certainly toxic, it is difficult to know how it applies to the common human condition of obesity.

IT'S INSULIN

THE ANSWER IS simpler once we understand hormonal obesity theory. Insulin is the major hormonal driver of weight gain. Insulin causes adult obesity. Insulin causes newborn obesity. Insulin causes infant obesity. Insulin causes childhood obesity. Where would an infant get high insulin levels? From his or her mother.

Dr. David Ludwig recently examined the relationship between the weights of 513,501 women and their 1,164,750 offspring.[9] Increased maternal weight gain is strongly associated with increased neonatal weight gain. Because both the mother and the fetus share the same blood supply, any hormonal imbalances, such as high insulin levels, are automatically and directly transmitted through the placenta from the mother to the growing fetus.

Fetal macrosomia is a term used for fetuses that are large for their gestational age. There are a number of risk factors, but chief among them are maternal gestational diabetes, maternal obesity and maternal weight gain. What do these conditions all have in common? High maternal levels of insulin. The high levels transmit to the developing fetus, resulting in one that is too large.

The logical consequence of too much insulin in the newborn is the development of insulin resistance, which leads to even higher levels of insulin in a classic vicious cycle. The high insulin levels produce obesity in the newborn as well as the six-month-old infant. The origins of both infant obesity and adult obesity are the same: insulin. These are not two separate diseases, but two sides of the same coin. Babies born to mothers with gestational diabetes mellitus have three times the risk of obesity and diabetes in later life, and one of the biggest risk factors for obesity in young adulthood is obesity in childhood.[10] Those who are obese in childhood have more than seventeen times the risk of obesity going into adulthood! Even large-for-gestational-age babies whose mothers do not have gestational diabetes are also at risk. They have double the risk of metabolic syndrome.

The sad but inescapable conclusion is that we are now passing on our obesity to our children. Why? Because we are now marinating our children in insulin starting in the womb, they develop more severe obesity sooner than ever before. Because obesity is time dependent and gets worse, fat babies become fat children. Fat children become fat adults. And fat adults have fat babies in turn, passing obesity on to the next generation.

What has really hampered our ability to combat childhood obesity, though, is a simple lack of understanding about the true causes of weight gain. A singular misguided focus on reducing caloric intake and increasing exercise led to government programs that have almost no chance of success. We didn't lack resources or willpower; we lacked knowledge of and a framework for understanding obesity.

SAME METHODS, SAME FAILURES

SEVERAL LARGE-SCALE STUDIES on prevention of childhood obesity were started in the late 1990s. The National Heart, Lung, and Blood Institute undertook the Pathways study at a cost of $20 million over eight years.[11] Dr. Benjamin Caballero, director of the Center for Human Nutrition at the Johns Hopkins Bloomberg School of Public Health, led this ambitious effort involving 1704 children in forty-one schools. Some schools received the special obesity-prevention program while other schools continued their standard program.

Low-income, native American children at risk for obesity and diabetes received both breakfast and lunch at the school cafeteria, where 'healthy' food lessons were immediately reinforced. Special exercise breaks were introduced in the middle of the school day. The specific nutritional goal was to reduce dietary fat to less than 30 per cent. In a nutshell, this was the same low-fat, low-calorie diet combined with increased exercise that had failed so miserably as a remedy for adult obesity.

Did the children learn how to eat a low-fat diet? Sure did. Dietary fat started at 34 per cent of calories and over the course of the study, fell to 27 per cent. Did they eat fewer calories? Sure did. The intervention group averaged 1892 calories per day compared to 2157 calories per day in the control group. Fantastic! The children were eating 265 fewer calories per day. They learned their lessons extremely well, eating fewer calories and less fat overall. Over the course of three years, calorie counters expected a loss of approximately 83 pounds! But did the children's weight actually change? *Not even by a little bit.*

Physical activity was no different between the two groups. Despite the increased physical education done in the schools, the total physical activity measured by accelerometer was not different—which should have been expected, given the known effect of compensation. Those children who were very active in school reduced their activity at home. Children relatively sedentary at school increased their activity once out of school.

This study was vitally important. The failure of the low-fat, low-calorie strategy should have prompted a search for more effective methods of controlling the scourge of childhood obesity. It should have prompted soul searching for the underlying cause of obesity and how to rationally treat it. So what happened?

The results were tabulated. The study was written. It was published in 2003 to thunderous... silence. Nobody wanted to hear the truth. The Eat Less, Move More approach, so adored by academic medicine, had failed yet again. It was easier to ignore the truth than to face it. And that's what happened.

Other studies confirmed these results. Dr. Philip Nader from the University of California San Diego randomized 5106 grade three to grade five students to education with 'healthy' food and increased exercise.[12] Fifty-six schools received the special program, and forty schools (the control group) did not. Once again, children receiving the extra indoctrination ate a lower-fat diet and retained this knowledge for years afterward. It was 'the largest school-based randomized trial ever conducted.' They ate less and exercised more. They just didn't lose any weight.

Obesity programs in community settings were similarly ineffective. The 2010 Memphis Girls Health Enrichment Multi-site Studies involved eight- to ten-year-old girls in a Memphis community center.[13] Group counseling encouraged subjects to 'reduce consumption of sugar-sweetened beverages and high-fat high caloric foods, increase intake of water, vegetables and fruit.' The message is very muddled here, but quite typical. Should we reduce sugar? Should we reduce fat?

Should we reduce calories? Should we eat more fruit? Should we eat more vegetables?

The program successfully reduced daily caloric intake from 1475 to 1373 at one year and further again to 1347 at two years. By contrast, the control group increased their daily caloric intake from 1379 to 1425 at two years. Did the girls lose weight? In a word, no. To add insult to injury, the body-fat percentage *increased* from 28 per cent to 32.2 per cent at the end of two years. A stunning failure for all involved, and yet another demonstration of the powerful calorie deception at work. Calories do not drive weight gain, and thus reducing them will not lead to weight loss.

But the persistently negative results were not enough to change ingrained beliefs. Both Drs. Caballero and Nader, rather than questioning their prior beliefs, felt that their treatments did not go far enough—a stance that is psychologically much, much easier to maintain.

While this appears absurd, when it comes to childhood obesity, we appear to have accepted the status quo. A low-fat, low-calorie diet combined with exercise is proven ineffective for weight loss—a finding that confirms our own common sense and observations. But instead of rethinking our failed strategy, we just continue, hoping against all hope that this time it will work.

SUCCESS AT LONG LAST

CONTRAST THAT TO the Australian Romp and Chomp study that ran from 2004 to 2008.[14] The program targeted almost 12,000 children aged zero to five years. Here again, daycare centers were divided into two groups. One group would continue their usual programs. The other intervention group received the Romp and Chomp educational initiative. But rather than giving multiple muddled health messages, the study's two major nutritional objectives were targeted and very specific:

1. To significantly decrease consumption of high-sugar drinks and promote the consumption of water and milk.
2. To significantly decrease consumption of energy-dense snacks and increase consumption of fruit and vegetables.

Rather than reducing fat and calories, the study reduced snacks and sugar. Similar to other programs, it tried to increase exercise and involve families as much as possible. But mostly, its methods were almost like your grandmother's advice to lose weight:

1. Cut down sugars and starches.
2. Stop snacking.

These strategies attack the worst offenders of insulin secretion and resistance. Snacks tend to be cookies, pretzels, crackers and other foods that are very high in refined carbohydrates, so reducing snacks reduced refined carbohydrate intake. Reducing sugar and refined carbohydrates will reduce insulin. Reducing snacking frequency prevents persistent high insulin levels, a key component of insulin resistance. These strategies lower insulin levels—the crucial, central problem of obesity. The program decreased consumption of packaged snacks and fruit juice (by approximately one-half cup daily). This study's results could not be more different from those of previous ones. Both the 2- and 3.5-year-old children showed significantly better weight reduction compared to the control group. The prevalence of obesity was reduced by 2- per cent to 3 per cent. Success at long last!

In southwest England, six schools launched a program called 'Ditch the Fizz.'[15] The single goal was to reduce soda drinking in children aged seven to eleven years. The program succeeded in reducing daily consumption by about 5 ounces (150 milliliters), which resulted in a decrease in obesity by 0.2 per cent. While it may seem trivial, obesity increased among the control group by a massive 7.5 per cent. Reducing the use of sugar-sweetened beverages is a highly effective method of preventing childhood obesity.

This program was effective because it contained a very specific message: reduce soda consumption. Other programs are too ambitious and too vague, and often, multiple mixed messages repeat in an endless loop. The importance of reducing intake of sugar-sweetened beverages can get lost in the cacophony.

WHAT YOUR GRANDMA SAID

WHILE STUDY AFTER study showed the failure of conventional weight-loss strategies, we plunged ahead with national exercise programs. We spent money and energy promoting exercise or building playgrounds in a misguided attempt to curb childhood obesity. When I grew up in the 1970s in Ontario, Canada, we had the ParticipACTION program, which was revived in 2007 at a cost of $5 million. ParticipACTION's explicit aim is increasing physical activity in children, with a tagline of 'Bring Back Play.' (Having watched my own children play exuberantly everywhere, I somehow doubt that 'play' is in danger of disappearing.) The original program that ran from the 1970s to the 1990s certainly failed to make a dent in the obesity crisis, but instead of burying these tired ideas, we resurrected them.

Michelle Obama launched the Let's Move! campaign with the ambitious goal of ending childhood obesity. Her strategy? Eat Less, Move More. Does she believe that this advice will work now, after forty years of uninterrupted failure? Insulin, not calories, causes weight gain. It's not (and never was) a matter of restricting calories. It's a matter of reducing insulin.

Despite the blunders, the news on childhood obesity is good. Recently, an unexpected ray of hope shone through the darkness. In 2014, the *Journal of the American Medical Association* reported that obesity rates for the age group two to five years had dropped by 43 per cent between 2003 and 2012.[16] There was no change in youth or adult rates of obesity. However, since childhood obesity is strongly linked to adult obesity, this is indeed very good news.

Some groups wasted no time in congratulating themselves on a job well done. They believe that their campaign of physical activity and caloric reduction has played a key role in this success. I don't buy it.

The answer is more straightforward. Consumption of added sugars steadily increased from 1977 along with obesity. By the late 1990s, increasing attention focused on the key role that sugar plays in weight gain. The irrefutable truth remained that sugar causes weight gain, with no redeeming nutritional qualities. Sugar intake began to fall in 2000, and after a five- to ten-year lag, so did obesity. We see this first in the youngest age group since they have had the least exposure to high insulin levels and therefore have less insulin resistance.

The most ironic part of this entire wretched episode is that we already knew the answers. The pediatrician Dr. Benjamin Spock wrote his classic bible of child rearing, *Baby and Child Care,* in 1946. For more than fifty years, it was the second-bestselling book in the world, after the Bible. Regarding childhood obesity, he writes, 'Rich desserts can be omitted without risk, and should be, by anyone who is obese and trying to reduce. The amount of plain, starchy foods (cereals, breads, potatoes) taken is what determines . . . how much (weight) they gain or lose.'[17]

This, of course, was exactly what Grandma would say. 'Cut back sugars and starchy foods. No snacking.' If only we had listened to Grandma instead of Big Brother.

PART FIVE

What's Wrong with Our Diet?

(14)

THE DEADLY EFFECTS OF FRUCTOSE

•

SUGAR IS FATTENING. This nutritional fact enjoys almost universal agreement. The 1977 *Dietary Guidelines for Americans* clearly warned against the dangers of excessive dietary sugar, but the message got lost in the anti-fat hysteria that followed. Dietary fat was the overwhelming concern of health-conscious shoppers, and the sugar content of food was ignored or forgotten. Bags of jellybeans and other candies were proudly proclaiming themselves to be fat free. The fact that they were virtually 100 per cent sugar didn't seem to bother anybody. Sugar consumption rose steadily from 1977 to 2000, paralleled by the rising obesity rates. Diabetes followed with a time lag of ten years.

IS SUGAR TOXIC?

THE WORST OFFENDER, by far, is the sugar-sweetened drink—soft drinks, sodas and, more recently, sweetened teas and juices. Soda is a $75 billion industry that had, until recently, seen nothing but good times. Per capita intake of sugar-sweetened drinks doubled in the

1970s. By the 1980s, sugar-sweetened drinks had become more popular than tap water. By 1998, Americans were drinking 56 gallons per year. By the year 2000, sugar-sweetened drinks provided 22 per cent of the sugar found in the American diet, compared to 16 per cent in 1970. No other food group even came close.[1]

Thereafter, sugar-sweetened drink relentlessly declined in popularity. From 2003 to 2013, soft-drink consumption in the United States dropped by close to 20 per cent.[2] Sweetened iced teas and sugary sports drinks have valiantly tried to take their place, but have been unable to block the winds of change. By 2014, Coca Cola had faced nine consecutive years of sales decline as health concerns about sugar mounted. Concerned with declining health and ballooning waistlines, people were less inclined to drink a toxic, sugary brew.

Sugar-sweetened drinks now face strong political opposition—from proposed soda taxes to the recent effort by New York mayor Michael Bloomberg to outlaw oversized beverages. Some of the problems, of course, are self-inflicted. Coca Cola spent decades convincing people to drink more soda. They were wildly successful, but at what cost? As the obesity crisis grew, the companies found themselves under increasing fire from all sides.

But the sugar pushers weren't so easily defeated. Knowing that they were fighting a losing battle in much of North America and Europe, they took aim at Asia to make up for lost profits. Asian sugar consumption is rising at almost 5 per cent per year, even as it has stabilized or fallen in North America.[3]

The result has been a diabetes catastrophe. In 2013, an estimated 11.6 per cent of Chinese adults have type 2 diabetes, eclipsing even the long-time champion: the U.S., at 11.3 per cent.[4] Since 2007, 22 million Chinese were newly diagnosed with diabetes—a number close to the population of Australia.[5] Things are even more shocking when you consider that only 1 per cent of Chinese had type 2 diabetes in 1980.[6] *In a single generation,* the diabetes rate rose by a horrifying 1160 per cent. Sugar, more than any other refined carbohydrate, seems to be particularly fattening and leads to type 2 diabetes.

Daily consumption of sugar-sweetened drinks not only carries a significant risk of weight gain, but also increases the risk of developing diabetes by 83 per cent compared to drinking less than one sugar-sweetened drink per month.[7] But is the culprit sugar or calories? Further research suggested that the prevalence of diabetes rose by 1.1 per cent for every extra 150 calories per person per day of sugar.[8] No other food group showed any significant relationship to diabetes. Diabetes correlates with sugar, not calories.

Sucrose, against all logic and common sense, had not been considered bad for diabetics. In 1983, Dr. J. Bantle, a prominent endocrinologist, asserted in the *New York Times* that 'the message is that diabetics may eat foods containing ordinary sugar, if they keep the amount of calories at the same constant level.'[9] The U.S. Food and Drug Administration (FDA) undertook a comprehensive review in 1986.[10] Citing more than 1000 references, the Sugars Task Force declared, 'there is no conclusive evidence on sugars that demonstrates a hazard.' In 1988, the FDA would reaffirm sugar as 'Generally Recognized as Safe.' In 1989, the National Academy of Sciences' report *Diet and Health: Implications for Reducing Chronic Disease* chimed in with the view that 'sugar consumption (by those with an adequate diet) has not been established as a risk factor for any chronic disease other than dental caries in humans.'[11]

Yes, cavities. There seemed to be no concern that eating more sugar would raise blood sugar. Even in 2014, the American Diabetes Association website stated that 'experts agree that you can substitute small amounts of sugar for other carbohydrate-containing foods into your meal plan.'[12]

Why sugar is so fattening? Sugar is sometimes considered 'empty calories,' containing few nutrients. It is also thought to make food more 'palatable' and 'rewarding,' causing overconsumption and obesity. But perhaps the fattening effect of sugar is due to its nature as a highly refined carbohydrate. It stimulates the production of insulin, which causes weight gain. But then again, most refined carbohydrates, such as rice and potatoes, do so too.

What was it specifically about sugar that seems to be particularly toxic? The INTERMAP study compared Asian and Western diets in the 1990s.[13] The Chinese, despite much higher intakes of refined carbohydrates, had far lower rates of diabetes. Part of reason for this advantage lies in the fact that their sugar consumption was much lower.

Sucrose differs from other carbohydrates in one important way. The problem? Fructose.

SUGAR BASICS

GLUCOSE, A SUGAR with the basic molecular structure of a six-sided ring, can be used by virtually every cell in the body. Glucose is the main sugar found in the blood and circulates throughout the body. In the brain, it is the preferred energy source. Muscle cells will greedily import glucose from the blood for a quick energy boost. Certain cells, such as red blood cells, can *only* use glucose for energy. Glucose can be stored in the body in various forms such as glycogen in the liver. If glucose stores run low, the liver can make new glucose via the gluconeogenesis process (literally meaning 'making new glucose').

Fructose, a sugar with the basic molecular structure of a five-sided ring, is found naturally in fruit. It is metabolized only in the liver and does not circulate in the blood. The brain, muscles and most other tissues cannot use fructose directly for energy. Eating fructose does not appreciably change the blood glucose level. Both glucose and fructose are single sugars, or monosaccharides.

Table sugar is called sucrose, and is composed of one molecule of glucose linked to one molecule of fructose. Sucrose is 50 per cent glucose and 50 per cent fructose. High-fructose corn syrup is composed of 55 per cent fructose and 45 per cent glucose. Carbohydrates are composed of sugars. When these carbohydrates contain a single sugar (monosaccharides) or two sugars (disaccharides), they are called simple carbohydrates. When many hundreds or even thousands of sugars are linked into long chains (polysaccharides), they are called complex carbohydrates.

However, it was recognized long ago that this classification provided little physiologically useful information, since it only differentiates based upon the chain length. It had previously been thought that complex carbohydrates were digested more slowly, causing less of a rise in blood sugar, but this is not true. For example, white bread, which is composed of complex carbohydrates, causes a very quick spike in blood sugar, almost as high as a sugar-sweetened drink.

Dr. David Jenkins reclassified foods according to their blood glucose effect in the early 1980s, which provided a useful comparison of the different carbohydrates. This pioneering work led to the development of the glycemic index. Glucose was given the value of 100, and all other foods are measured against this yardstick. Bread, both whole wheat and white, has a glycemic index of 73, comparable to Coca-Cola, which has a value of 63. Peanuts, on the other hand, have a very low value of 7.

There is an unspoken assumption that most of the negative effects of carbohydrates are due to their effect on blood glucose, but this idea is not necessarily true. Fructose, for example, has an extremely low glycemic index. Furthermore, it should be noted that the glycemic index measures blood glucose, not blood insulin levels.

FRUCTOSE: THE MOST DANGEROUS SUGAR

WHERE DOES FRUCTOSE fit in? Fructose does not raise the blood glucose appreciably, yet is even more strongly linked to obesity and diabetes than glucose. From a nutritional standpoint, neither fructose nor glucose contains essential nutrients. As a sweetener, both are similar. Yet fructose seems particularly malevolent to human health.

Fructose was previously considered a benign sweetener because of its low glycemic index. Fructose is found naturally in fruits, and is the sweetest naturally occurring carbohydrate. What could be wrong with that?

The problem, as often is the case, is a matter of scale. Natural fruit consumption contributed only small amounts of fructose to our diet, in the range of 15 to 20 grams per day. Things began to change with the development of high-fructose corn syrup. Fructose consumption steadily rose until the year 2000, when it peaked at 9 per cent of total calories. Adolescents in particular were heavy users of fructose at 72.8 grams per day.[14]

High-fructose corn syrup was developed in the 1960s as a liquid equivalent of sucrose. Sucrose was processed from sugar cane and sugar beets. While not exactly expensive, it wasn't exactly cheap. High-fructose corn syrup, however, could be processed from the river of cheap corn that was flowing out of the American midwest—and that was the decisive factor in favor of high-fructose corn syrup. It was cheap.

In processed food, high-fructose corn syrup found a natural partner. As a liquid, it could easily be incorporated into processed food. But its advantages didn't stop there. Just consider that it

- is sweeter than glucose,
- prevents freezer burn,
- helps browning,
- mixes easily,
- extends shelf life,
- keeps breads soft and
- has a low glycemic index.

Soon, high-fructose corn syrup found its way into almost every processed food. Pizza sauce, soups, breads, cookies, cakes, ketchup, sauces—you name it, it probably contained high-fructose corn syrup. It was cheap, and big food companies cared about cost more than anything else in the world. Food manufacturers raced to use high-fructose corn syrup at every opportunity.

Fructose has an extremely low glycemic index. Sucrose and high-fructose corn syrup, with roughly 55 per cent fructose, have significantly better glycemic-index measures than glucose. Furthermore, fructose produces only a mild rise in insulin levels compared to glucose, which led many people to regard fructose as a more benign form of sweetener. Fructose is also the main sugar in fruit, adding to its halo. An all-natural fruit sugar that doesn't raise blood sugars? Sounded pretty healthy. A wolf in sheep's clothing? You bet your life. The difference between glucose and fructose will very literally kill you.

The tide began to turn in 2004 when Dr. George Bray from the Pennington Biomedical Research Center of Louisiana State University showed that the increase in obesity closely mirrored the rise in use of high-fructose corn syrup. (See Figure 14.1.[15])In the public consciousness, high-fructose corn syrup developed as a major health issue. Others correctly pointed out that high-fructose corn syrup use increased in proportion to the decreased use of sucrose. The rise in obesity really mirrored the increase in total fructose consumption, whether the fructose came from sucrose or from high-fructose corn syrup.

But why was fructose so bad?

Figure 14.1. Obesity rates have risen in proportion to high-fructose corn syrup intake.

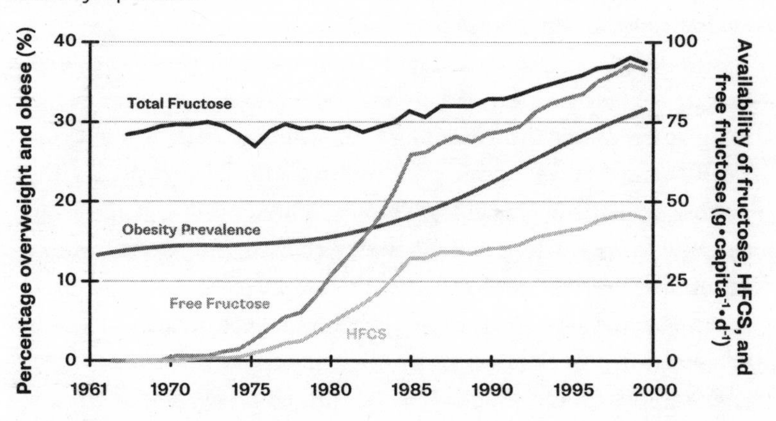

FRUCTOSE METABOLISM

AS THE DANGERS of dietary fructose received increased scrutiny, researchers scrambled to investigate. Glucose and fructose differ in many significant ways. Whereas almost every cell in the body can use glucose for energy, no cell has the ability to use fructose. Where glucose requires insulin for maximal absorption, fructose does not. Once inside the body, only the liver can metabolize fructose. Where glucose can be dispersed throughout the body for use as energy, fructose is targeted like a guided missile to the liver.

Excessive fructose puts significant pressure on the liver since other organs cannot help. It is the difference between pressing down with a hammer and pressing down with a needlepoint: much less pressure is needed if it is all directed onto a single point.

At the liver, fructose is rapidly metabolized into glucose, lactose and glycogen. The body handles excess glucose consumption through several well-defined metabolic pathways, such as glycogen storage and de novo lipogenesis (creation of new fat). No such system is present for fructose. The more you eat, the more you metabolize. The bottom line is that excess fructose is changed into fat in the liver. High levels of fructose will cause fatty liver. *Fatty liver is absolutely crucial to the development of insulin resistance in the liver.*

That fructose directly causes insulin resistance was discovered long ago. As far back as 1980, experiments proved that fructose (but not glucose) caused the development of insulin resistance in humans.[16] Healthy subjects were given an extra 1000 calories per day of either glucose or fructose. The glucose group showed no change in insulin sensitivity. The fructose group, however, showed a 25 per cent worsening of their insulin sensitivity—after just seven days!

A 2009 study showed that pre-diabetes could be induced in healthy volunteers in only eight weeks. Healthy subjects ate 25 per cent of their daily calories as Kool-Aid sweetened with either glucose or fructose. While this seems high, many people consume this high proportion of sugar in their diets.[17] With its low glycemic index, fructose raised blood glucose much less.

The fructose, but not the glucose group, *developed pre-diabetes by eight weeks.* Insulin levels as well as measures of insulin resistance were significantly higher in the fructose group.

So only six days of excess fructose will cause insulin resistance. By eight weeks, pre-diabetes is establishing a beachhead. What happens after *decades* of high fructose consumption? *Fructose overconsumption leads directly to insulin resistance.*

MECHANISMS

INSULIN IS NORMALLY released when we eat. It directs some of the incoming glucose to be used as energy and some to be stored for later use. In the short term, glucose is stored as glycogen in the liver, but the liver's storage space for glycogen is limited. Once it's full, excess glucose is stored as fat: that is, the liver begins manufacturing fat from glucose through de novo lipogenesis.

After the meal, as insulin levels fall, this process reverses. With no food energy coming in, stored food energy must be retrieved. Glycogen and fat stores in the liver are turned back into glucose and distributed to the rest of the body for energy. The liver acts like a balloon. As energy comes in, it fills up. As energy is needed, it deflates. Balancing feeding and fasting periods over a day ensures that no net fat is gained or lost.

But what happens if the liver is already crammed full of fat? Insulin then tries to force more fat and sugar into the liver, even though it's already full of fat and sugar. Just as it is more difficult to inflate a fully inflated balloon, insulin has more difficulty trying to shove more fat into a fatty liver. It takes higher and higher levels of insulin to move the same amount of food energy into a fatty liver. The body is now resistant to the efforts of insulin, since normal levels will not be enough to push sugar into the liver. Voilà—insulin resistance in the liver.

The liver, like an overinflated balloon, will try to expel the sugar back into circulation, so continuously high insulin levels are also required to keep it bottled up in the liver. If insulin levels start to drop, the stored fat and sugar comes whooshing out. To compensate, the body keeps raising its insulin levels.

Thus, insulin resistance leads to higher insulin levels. High insulin levels encourage more storage of sugar and fat in the liver, which causes even more over-cramming of fat in the already fatty liver, causing more insulin resistance—a classic vicious cycle.

Sucrose, a fifty-fifty mix of glucose and fructose, therefore plays a dual role in obesity. Glucose is a refined carbohydrate that directly stimulates insulin. Fructose overconsumption causes fatty liver, which directly produces insulin resistance. Over the longer term, insulin resistance also leads to increased insulin levels, which then feeds back to increase insulin resistance.

Sucrose stimulates insulin production *both* in the short term and in the long term. In this way, sucrose is twice as bad as glucose. The effect of glucose is obvious in the glycemic index, but the effect of fructose is completely hidden. This fact misled scientists to downplay the role of sucrose in obesity.

But the uniquely fattening effect of sugar has finally been recognized. Cutting back on sugars and sweets has always been the first step in weight reduction in virtually all diets throughout history. Sugars are not simply empty calories or refined carbohydrates. They are far more dangerous than that, as they stimulate *both* insulin and insulin resistance.

The extra fattening effect of sugar is due to the stimulation of insulin resistance from fructose, which festers for years or even decades before it becomes obvious. Short-term feeding studies completely miss this effect, as evidenced by a recent systemic analysis. Analyzing many studies lasting less a week, it concluded that fructose shows no special effect outside of its calories.[18] That's analogous to analyzing smoking studies lasting several weeks and concluding that smoking does not cause lung cancer. Sugar's effects, as well as obesity, develop over *decades,* not days.

This explains the apparent paradox of the Asian rice eater. The INTERMAP studies of the 1990s found that the Chinese were eating very high amounts of white rice, but suffered little obesity. The key was that their sucrose consumption was extremely low, which minimized the development of insulin resistance.

Once their sucrose consumption started to increase, they began to develop insulin resistance. Combined with their original high carbohydrate intake (white rice), this was a recipe for the diabetes disaster they are facing right now.

WHAT TO DO

IF YOU WANT to avoid weight gain, remove all added sugars from your diet. On this, at least, everybody can agree. Don't replace them with artificial sweeteners—as we'll see in the next chapter, those are equally bad.

Despite all the doom and gloom of the obesity epidemic, I am actually quite optimistic that we may have turned the corner. At last, the evidence is accumulating. The relentless increase of obesity in the United States has recently started to slow, and in some states may, for the first time, begun to decline.[19] According the Centers for Disease Control, the rate of new cases of type 2 diabetes is also starting to slow.[20] The reduction of dietary sugars plays no small role in this victory.

(15)

THE DIET SODA DELUSION

•

ON A WARM June night in 1879, the Russian chemist Constantin Fahlberg sat down to dinner and bit into a remarkably sweet roll of bread. What was remarkable was that no sugar was used to make it. Earlier that day, while working on coal-tar derivatives in the laboratory, an extraordinarily sweet experimental compound had spilled all over his hands, and then made its way into the rolls. Rushing back to the laboratory, he tasted everything in sight. He had just discovered saccharin, the world's first artificial sweetener.

THE SEARCH FOR SWEETENERS

ORIGINALLY SYNTHESIZED AS a drink additive for diabetics, saccharin's popularity slowly spread,[1] and eventually other sweet, low-calorie compounds were synthesized.

Cyclamate was discovered in 1937, but later removed from use in the United States in 1970 due to concerns about bladder cancer. Aspartame (NutraSweet), was discovered in 1965. Approximately 200 times sweeter than sucrose, aspartame is one of the most notorious of sweeteners, due to its cancer-causing potential in animals. Nevertheless, it gained approval for use in 1981. Aspartame's popularity has since been eclipsed by acesulfame potassium, followed by the current champion, sucralose. Diet soda is the most obvious source in our diet of these chemicals, but yogurts, snack bars, breakfast cereals and many other 'sugar-free' processed foods also contain them.

Diet drinks contain very few calories and no sugar. Therefore, replacing a regular soft drink with a diet soda *seems* like a good way to reduce sugar intake and help shed some pounds. With the increasing health concerns around excess sugar, food manufacturers responded by releasing an estimated 6000 new artificially sweetened products. The intake of artificial sweeteners has increased markedly in the U.S. population (see Figure 15.1)[2] with 20 per cent to 25 per cent of American adults routinely ingesting these chemicals, mostly in beverages.

Figure 15.1. Per capita consumption of artificial sweeteners increased more than 12-fold between 1965 and 2004.

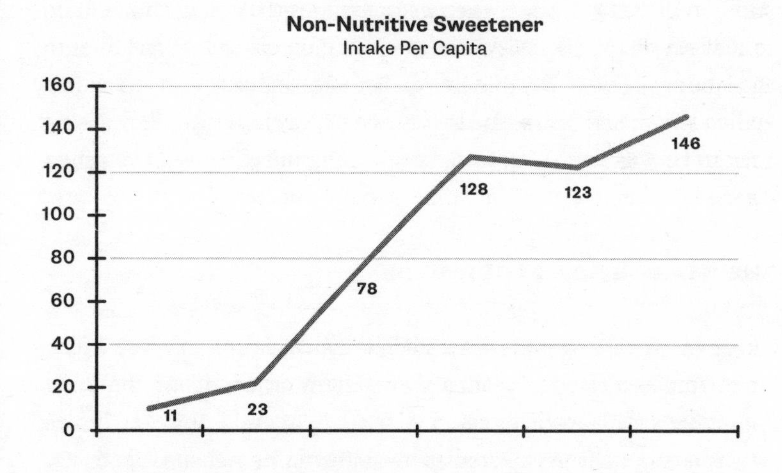

From humble beginnings in 1960 to the year 2000, the consumption of diet soda has increased by more than 400 per cent. Diet Coke has long been the second most popular soft drink, just behind regular

Coca Cola. In 2010, diet drinks made up 42 per cent of Coca Cola's sales in the United States. Despite initial enthusiasm, though, the use of artificial sweeteners has recently leveled off, primarily due to safety concerns. Surveys indicate that 64 per cent of respondents had some health concerns about artificial sweeteners, with 44 per cent making a deliberate effort to reduce their intake or avoid them altogether.[3]

And so the search has been on for more 'natural' low-calorie sweeteners. Agave nectar enjoyed a brief surge of popularity. Agave nectar is processed from the agave plant, which grows in the southwest regions of the United States, Mexico and parts of South America. Agave was felt to be a healthier alternative to sugar due to its lower glycemic index. Dr. Mehmet Oz, a cardiologist popular on American television, briefly touted the health benefits of agave nectar before reversing his stance when he realized it was mostly fructose (80 per cent).[4] Agave's low glycemic index was simply due to its high fructose content.

The next big thing to hit the market was stevia. Stevia is extracted from the leaves of *Stevia rebaudiana,* a plant that is native to South America. It has 300 times the sweetness of regular sugar and a minimal effect on glucose. Widely used in Japan since 1970, it has recently become available in North America. Both agave nectar and sweeteners derived from Stevia are highly processed. In that regard, they are not any better than sugar itself—a natural compound derived from sugar beets.

THE SEARCH FOR PROOF

IN 2012, BOTH the American Diabetes Association and the American Heart Association issued a joint statement endorsing the use of low-calorie sweeteners to aid in losing weight and improving health.[5] The American Diabetes Association states on its website, 'Foods and drinks that use artificial sweeteners are another option that may help curb your cravings for something sweet.'[6] But the evidence is surprisingly scarce.

Presuming that artificial low-calorie sweeteners are beneficial presents an immediate and obvious problem. Per capita consumption of diet foods has skyrocketed in recent decades. If diet drinks substantially reduce obesity or diabetes, why did these two epidemics continue unabated? The only logical conclusion is that diet drinks don't really help.

There are substantial epidemiologic studies to back that up. The American Cancer Society conducted a survey of 78,694 women, hoping to find that artificial sweeteners had a beneficial effect on weight. Instead, the survey showed exactly the opposite. After adjustment for initial weight, over a one-year period, those using artificial sweeteners were significantly *more* likely to gain weight, although the weight gain itself was relatively modest (less than 2 pounds).[7]

Dr. Sharon Fowler, from the University of Texas Health Sciences Center at San Antonio, in the 2008 San Antonio Heart Study prospectively studied 5158 adults over eight years.[8] She found that instead of reducing the obesity, diet beverages substantially increased the risk of it by a mind-bending 47 per cent. She writes, 'These findings raise the question whether [artificial sweetener] use might be fueling—rather than fighting—our escalating obesity epidemic.'

The bad news for diet soda kept rolling in. Over the ten years of the Northern Manhattan Study, Dr. Hannah Gardener from the University of Miami found in 2012 that drinking diet soda was associated with a 43 per cent increase in risk of vascular events (strokes and heart attacks).[9] The 2008 Atherosclerosis Risk in Communities Study (ARIC)[10] found a 34 per cent increased incidence of metabolic syndrome in diet soda users, which is consistent with data from the 2007 Framingham Heart Study,[11] which showed a 50 per cent higher incidence of metabolic syndrome. In 2014, Dr. Ankur Vyas from the University of Iowa Hospitals and Clinics presented a study following 59,614 women over 8.7 years in the Women's Health Initiative Observational Study.[12] The study found a 30 per cent increase risk of cardiovascular events (heart attacks and strokes) in those drinking two or more diet drinks daily.

The benefits for heart attack, stroke, diabetes and metabolic syndrome were similarly elusive. Artificial sweeteners are not *good.* They are *bad.* Very bad.

Despite reducing sugar, diet sodas do not reduce the risk of obesity, metabolic syndrome, strokes or heart attacks. But why? Because it is insulin, not calories, that ultimately drives obesity and metabolic syndrome.

The important question is this: Do artificial sweeteners increase insulin levels? Sucralose raises insulin by 20 per cent, despite the fact that it contains no calories and no sugar.[13] This insulin-raising effect has also been shown for other artificial sweeteners, including the 'natural' sweetener stevia. Despite having a minimal effect on blood sugars, both aspartame and stevia raised insulin levels *higher even than table sugar.*[14] Artificial sweeteners that raise insulin should be expected to be harmful, not beneficial. Artificial sweeteners may decrease calories and sugar, but not insulin. Yet it is insulin that drives weight gain and diabetes.

Artificial sweeteners may also cause harm by increasing cravings. The brain may perceive an incomplete sense of reward by sensing sweetness without calories, which may then cause overcompensation and increased appetite and cravings.[15] Functional MRI studies show that glucose activates the brain's reward centers fully—but not sucralose.[16] The incomplete activation could stimulate cravings for sweet food to fully activate the reward centers. In other words, you may be developing a habit of eating sweet foods, leading to overeating. Indeed, most controlled trials show that there is no reduction in caloric intake with the use of artificial sweeteners.[17]

The strongest proof of failure comes from two recent randomized trials. Dr. David Ludwig from Harvard randomly divided two groups of overweight adolescents.[18] One group was given water and diet drinks to consume while the control group continued with their usual drinks. At the end of two years, the diet soda group was consuming far less sugar than the control group. That's good—but that is not our question.

Does drinking diet soda make any difference to adolescent obesity? The short answer is no. *There was no significant weight difference between the two groups.*

Another shorter-term study involving 163 obese women randomized to aspartame did not show improved weight loss over nineteen weeks.[19] But one trial involving 641 normal-weight children did find a statistically significant weight loss associated with the use of artificial sweeteners.[20] However, the difference was not as dramatic as hoped. At the end of eighteen months, there was only a 1-pound difference between the artificial sweetener group and the control group.

Conflicting reports such as these often generate confusion within nutritional science. One study will show a benefit and another study will show the exact opposite. Generally, the deciding factor is who paid for the study. Researchers looked at seventeen different reviews of sugar-sweetened drinks and weight gain.[21] A full 83.3 per cent of studies sponsored by food companies did not show a relationship between sugar-sweetened drinks and weight gain. But independently funded studies showed the exact opposite—83.3 per cent showed a strong relationship between sugar-sweetened drinks and weight gain.

THE AWFUL TRUTH

THE FINAL ARBITER, therefore, must be common sense. Reducing dietary sugars is certainly beneficial. But that doesn't mean that replacing sugar with completely artificial, manmade chemicals of dubious safety is a good idea. Some pesticides and herbicides are also considered safe for human consumption. However, we shouldn't be going out of our way to eat more of them.

Caloric reduction is the main advantage of artificial sweeteners. But it is not calories that drives obesity; it's insulin. Since artificial sweeteners also raise insulin levels, there is no benefit to using them. Eating chemicals that are not foods (such as aspartame, sucralose or acesulfam potassium) is not a good idea. They are synthesized in large chemical vats and added to foods because they happen to be sweet and not kill you. Small amounts of glue won't kill you either. That doesn't mean we should be eating it.

The bottom line is that these chemicals do not help you lose weight and may actually cause you to gain it. They may cause cravings that induce overeating of sweet foods. And continually eating sweet foods, even if they have no calories, may lead us to crave other sweet foods.

The randomized trials confirm our own personal experience and common sense. Yes, drinking diet soda will reduce sugar intake. But no, it will not help reduce your weight. This, of course, you probably already knew. Consider all the people you see drinking diet sodas. Do you know *anybody at all* who said that drinking diet soda made him or her lose a lot of weight?

Anybody at all?

(16)

CARBOHYDRATES AND PROTECTIVE FIBER

•

CONTROVERSY SURROUNDS THE humble carbohydrate. Is it good or bad? From the mid 1950s to the 1990s, they were the good guys, the heroes. Low in fat, they were supposed to be our salvation from the 'epidemic' of heart disease. Then, the Atkins onslaught of the late 1990s recast them in the role of dietary villain. Many advocates avoided all carbohydrates—even vegetables and fruits. So, are carbohydrates good or bad?

Insulin and insulin resistance drive obesity. *Refined* carbohydrates, such as white sugar and white flour, cause the greatest increase in insulin levels. These foods are quite fattening, but that doesn't necessarily mean that *all* carbohydrates are similarly bad. 'Good' carbohydrates (whole fruits and vegetables) are substantially different from 'bad' (sugar and flour). Broccoli will likely not make you fat, no matter how much you eat. But eating even modest amounts of sugar can certainly cause weight gain. Yet both are carbohydrates. How do we distinguish the two?

GLYCEMIC INDEX AND GLYCEMIC LOAD

DR. DAVID JENKINS of the University of Toronto began to tackle this problem in 1981 with the glycemic index. Foods were ranked according to their ability to raise glucose levels. Since dietary protein and fat did not raise blood glucose appreciably, they were excluded from the glycemic index, which measures only carbohydrate-containing foods. For those foods, glycemic index and insulin-stimulating effect are closely correlated.

The glycemic index uses identical 50-gram portions of carbohydrate. For example, you might take foods such as carrots, watermelon, apples, bread, pancakes, a candy bar and oatmeal, measure out a portion of each to contain 50 grams of carbohydrate, then measure the effect on blood glucose. Then you compare the foods against the reference standard—glucose—which is assigned a value of 100.

However, a standard serving of food may not contain 50 grams of carbohydrate. For example, watermelon has a very high glycemic index of 72, but contains only 5 per cent carbohydrate by weight. Most of watermelon's weight is water. You would need to eat 1 kilogram (2.2 pounds!) of watermelon to get 50 grams of carbohydrate—far in excess of what a person would eat at one sitting. A corn tortilla, though, has a glycemic index of 52. The tortilla is 48 per cent carbohydrate by weight, so you would only have to eat 104 grams of the tortilla (close to what a person would reasonably eat during a meal) to get 50 grams of carbohydrate.

The glycemic load index attempts to correct this distortion by adjusting for serving size. Watermelon turns out to have a very low glycemic load of 5 while the corn tortilla still ranks high at 25. But whether you use the glycemic index or glycemic load, you'll find there is a clear distinction between refined carbohydrates and unrefined traditional foods. Western refined foods have a very high glycemic index and glycemic load scores. Traditional whole foods have low glycemic load scores, despite containing similar amounts of carbohydrate—an essential distinguishing feature. (See Figure 16.1[1].) Carbohydrates are not inherently fattening. *Their toxicity lies in way they are processed.*

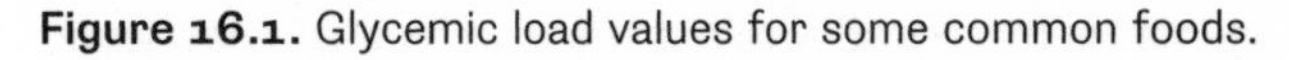

Figure 16.1. Glycemic load values for some common foods.

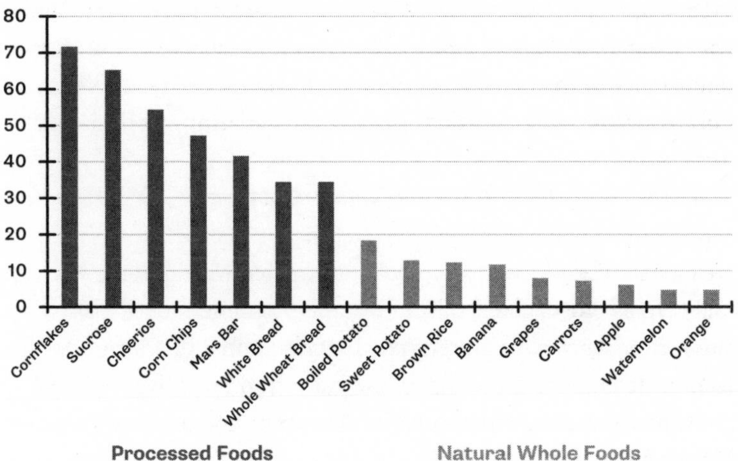

Refining significantly increases the glycemic index by purifying and concentrating the carbohydrate. Removal of fat, fiber and protein means that the carbohydrate can be digested and absorbed very quickly. In the example of wheat, modern machine milling, which has almost completely replaced the traditional stone milling, pulverizes the wheat into the very fine white powder we know as flour. Cocaine users will know that very fine powders are absorbed into the bloodstream much faster than coarse grains—that's what allows for higher 'highs,' both for cocaine and for glucose. Refined wheat causes our glucose levels to spike. Insulin levels follow.

Second, refining encourages overconsumption. For example, making a glass of orange juice may require four or five oranges. It is very easy to drink a glass of juice, but eating five oranges is not so easy. By removing everything other than the carbohydrate, we tend to overconsume what is left. If we had to eat all the fiber and bulk associated with five oranges, we might think twice about it. The same applies to grains and vegetables.

The problem is one of balance. Our bodies have adapted to the balance of nutrients in natural food. By refining foods and only consuming a certain component, the balance is entirely destroyed. People have been eating unrefined carbohydrates for thousands of years without obesity or diabetes. What's changed, and recently too, is that we now predominantly eat *refined grains* as our carbohydrate of choice.

WHEAT: THE WEST'S GRAIN OF CHOICE

WHEAT HAS LONG been a symbol of nutrition. Wheat, along with rice and corn, is one of the first domesticated foods in human history. Yet these days, what with gluten sensitivity and obesity, wheat does not have a friend to call its own. But how can wheat possibly be so bad?

As discussed in chapter 9, wheat has been cultivated since ancient times. But by the 1950s, Malthusian concerns of overpopulation and worldwide famine arose again. Norman Borlaug, who would later win the Nobel Peace Prize, began experimentating with higher-yield wheat varieties, and thus was born the dwarf-wheat variety.

Today, an estimated 99 per cent of all wheat grown worldwide is dwarf or semi-dwarf varieties. But where Dr. Borlaug bred naturally occurring strains together, successors quickly turned to new technologies to enhance mutations. The new varieties of wheat were not tested for safety, but were merely assumed to be safe in this new atomic age.

It is clear that the dwarf wheat varieties of today are not the same as those fifty years ago. The Broadbalk Wheat Experiment documented the change in nutritional content over the last half century.[2] Even as grain yields skyrocketed during the Green Revolution, the micronutrient content plummeted. Today's wheat is simply not as nutritious as in previous generations. That surely cannot be good news.

Another clue to wheat's changing character is the enormous increase in the prevalence of celiac disease, which is a reaction against gluten protein that damages the small intestine. Wheat is by far the predominant source of gluten in the Western diet, often by a factor of 100 or more. By comparing archived blood samples from Air Force men over a period of fifty years, researchers discovered that the prevalence of celiac disease appears to have quadrupled.[3] Could this be a result of new wheat varieties? This question has not yet been satisfactorily answered, but the possibility is troubling.

Processing methods have changed significantly over the centuries. Wheat berries were traditionally ground by large millstones powered by animals or humans. The modern flourmill has replaced traditional stone grinding. The bran, middlings, germ and oils are efficiently and completely removed, leaving the pure white starch. Most of the vitamins, proteins, fiber and fats are removed along with the outer hull and bran. The flour is ground to such a fine dust that its absorption by the intestine is extremely rapid. The increased rate of glucose absorption amplifies the insulin effect. Whole wheat and whole grain flours retain some of the bran and germ, but suffer from the same problem of rapid absorption.

Starches are hundred of sugars all linked together. Most (75 per cent) of the starch found in white flour is organized into branched chains called amylopectin; the remainder into amylose. There are several classes of amylopectin: A, B and C. Legumes are particularly rich in amylopectin C, which is very poorly digested. As the undigested carbohydrate moves through the colon, gut flora produces gas causing the familiar 'tooting' of the bean eater. While beans and legumes are very high in carbohydrate, much of it is not absorbed.

Amylopectin B, found in bananas and potatoes, is intermediate in terms of absorption. The most easily digested is amylopectin A found in—you guessed it—wheat. Wheat is converted to glucose more efficiently than virtually any other starch.

However, despite all the concerns discussed in this chapter, observational studies consistently demonstrate that whole grains are *protective* against obesity and diabetes. Where is the disconnection? The answer, here is fiber.

THE BENEFITS OF FIBER

FIBER IS THE non-digestible part of food, usually of a carbohydrate. Common types of fiber include cellulose, hemicellulose, pectins, beta-glucans, fructans and gums.

Fiber is classified as soluble or insoluble based on whether it is dissolvable in water. Beans, oat bran, avocado and berries are good sources of soluble fiber. Whole grains, wheat germ, beans, flax seeds, leafy vegetables and nuts are good sources of insoluble fiber. Fiber can also classified as fermentable or non-fermentable. Normal bacteria residing in the large intestine have the ability to ferment certain undigested fiber into the short-chain fatty acids acetate, butyrate and propionate, which can be used as an energy source. They may also have other beneficial hormonal effects, including the decreased output of glucose from the liver.[4] Generally, soluble fiber is more fermentable than insoluble.

Fiber has multiple purported mechanisms of health, but the importance of each is largely unknown. High-fiber foods require more chewing, which may help to reduce food intake. Horace Fletcher (1849–1919) believed strongly that chewing every bite of food 100 times would cure obesity and increase muscle strength. Doing so helped him lose 40 pounds (18 kilograms), and 'Fletcherizing' became a popular weight-loss method in the early twentieth century.

Fiber may decrease the palatability of food and thus reduce food intake. Fiber bulks up foods and decreases its energy density. Soluble fiber absorbs water to form a gel, further increasing its volume. This effect helps fill the stomach, which increases satiety. (Stomach distention may signal a sensation of fullness or satiety through the vagus nerve.) Increased bulk may also mean that the stomach takes more time to empty. Therefore, after meals rich in fiber, blood glucose and insulin levels are slower to rise. In some studies, half the variance of the glucose response to starchy foods depended on their fiber content.[5]

In the large intestine, the increased stool bulk may lead to increased caloric excretion. On the flip side, fermentation in the colon may produce short-chain fatty acids.[6] Roughly 40 per cent of dietary fiber may be metabolized in this way. One study demonstrated that a low-fiber diet resulted in 8 per cent higher caloric absorption.[7] In short, fiber may decrease food intake, slow down food's absorption in the stomach and small intestine, then help it exit quickly through the large intestines—all of which are potentially beneficial in treating obesity.

Fiber intake has fallen considerably over the centuries. In Paleolithic diets, it was estimated to be 77 to 120 grams per day.[8] Traditional diets are estimated to have 50 grams per day of dietary fiber.[9] By contrast, modern American diets contain as little as 15 grams per day.[10] Indeed, even the American Heart Association's *Dietary Guidelines for Healthy North American Adults* only recommends 25 to 30 grams per day.[11] However, removal of dietary fiber is a key component of food processing. And improving the texture, taste and consumption of foods directly increases food companies' profits.

Fiber came to public attention in the 1970s, and by 1977, the new *Dietary Guidelines* recommended we 'eat foods with adequate starch and fiber.' With that, fiber was enshrined in the pantheon of conventional nutritional wisdom. Fiber was good for you. But it was difficult to show exactly *how* it was good for you.

At first, it was believed that high fiber intake reduced colon cancer. The subsequent studies proved to be a bitter disappointment. The 1999 prospective Nurses' Health Study followed 88,757 women over sixteen years, and found no significant benefit in reducing colon cancer risk.[12] Similarly, a randomized study from 2000 of high fiber intake failed to demonstrate any reduction in precancerous lesions called adenomas.[13]

If fiber wasn't helpful in reducing cancer, perhaps fiber might be beneficial in reducing heart disease. The 1989 Diet and Reinfarction Trial randomized 2033 men after their first heart attack to three different diets.[14] To the researchers' astonishment, the American Heart Association's low-fat diet did not seem to reduce risk at all. What about a high-fiber diet? No benefit.

The Mediterranean diet (which is high fat), on the other hand, was beneficial, as Dr. Ancel Keys had suspected years ago. Recent trials such as the PREDIMED confirm the benefits of eating more natural fats such as nuts and olive oil.[15] So eating *more* fat is beneficial.

But it was difficult to shake the feeling that somehow, fiber was *good.* Many correlation studies, including the Pima and native Canadians, associate lower body mass index with higher fiber intake.[16, 17, 18] More recently, the ten-year observational CARDIA Study[19] found that those eating the most fiber were the least likely to gain weight. Short-term studies show that fiber increases satiety, reduces hunger and decreases caloric intake.[20] Randomized trials of fiber supplements show relatively modest weight-loss effects, with a mean weight loss of 2.9 to 4.2 pounds (1.3 to 1.9 kilograms) over a period of up to twelve months. Longer-term studies are not available.

FIBER: THE ANTI-NUTRIENT

WHEN WE CONSIDER the nutritional benefits of food, we typically consider the vitamins, minerals and nutrients contained. We think about components in the food that nourish the body. Such is not the case for fiber. The key to understanding fiber's effect is to realize that it is not as a nutrient, but as an *anti*-nutrient—where its benefit lies. Fiber has the ability to *reduce* absorption and digestion. Fiber subtracts rather than adds. In the case of sugars and insulin, this is good. Soluble fiber *reduces* carbohydrate absorption, which in turn reduces blood glucose and insulin levels.

In one study,[21] type 2 diabetic patients were split into two groups and given standardized liquid meals, one control group and the other with added fiber. The group that received liquid meals with added fiber reduced both the glucose and the insulin peaks, despite the fact that the two groups consumed exactly the same amount of carbohydrates and calories. Because insulin is the main driver of obesity and diabetes, its reduction is beneficial. In essence, fiber acts as a sort of 'antidote' to the carbohydrate—which, in this analogy, is the poison. (Carbohydrates, even sugar, are not literally poisonous, but the comparison is useful to understand fiber's effect.)

It is no coincidence that virtually all plant foods, in their natural, unrefined state, contain fiber. Mother Nature has pre-packaged the 'antidote' with the 'poison.' Thus, traditional societies may follow diets high in carbohydrates without evidence of obesity or type 2 diabetes. The one critical difference is that the carbohydrates consumed by traditional societies are unrefined and unprocessed, resulting in very high fiber intake.

Western diets are characterized by one defining feature—and it's not the amount of fat, salt, carbohydrate or protein. It's the high amount of *processing* of foods. Consider traditional Asian markets, full of fresh meats and vegetables. Many people in Asian cultures buy fresh food daily, so processing it to extend shelf life is neither necessary nor welcome. By contrast, North American supermarkets have aisles overflowing with boxed, processed foods. Several more aisles are dedicated to processed frozen foods. North Americans will buy groceries for weeks or even months at a time. The large-volume retailer Costco, for example, depends on this practice.

Fiber and fat, key ingredients, are removed in the refining process: fiber, to change the texture and make food taste 'better,' and natural fats, to extend shelf life, since fats tend to go rancid with time. And so we ingest the 'poison' without the 'antidote'—the protective effects of fiber is removed from much of our food.

Where whole, unprocessed carbohydrates virtually always contain fiber, dietary proteins and fats contain almost none. Our bodies have evolved to digest these foods without the need for fiber: without the 'poison,' the 'antidote' is unnecessary. Here again, Mother Nature has proven herself to be far wiser than us.

Removing protein and fat in the diet may lead to overconsumption. There are natural satiety hormones (peptide YY, cholecystokinin) that respond to protein and fat. Eating pure carbohydrate does not activate these systems and leads to overconsumption (the second-stomach phenomenon).

Natural foods have a balance of nutrients and fiber that, over millennia, we have evolved to consume. The problem is not with each specific component of the food, but rather the overall balance. For example, suppose we bake a cake with a balance of butter, eggs, flour and sugar. Now we decide to remove completely the flour and double the eggs instead. The cake tastes horrible. Eggs are not necessarily bad. Flour is not necessarily good, but the balance is off. The same holds true for carbohydrates. The entire package of unrefined carbohydrates, with fiber, fat, protein and carbohydrate is not necessarily bad. But removing everything except the carbohydrate destroys the delicate balance and makes it harmful to human health.

FIBER AND TYPE 2 DIABETES

BOTH OBESITY AND type 2 diabetes are diseases caused by excessive insulin. Insulin resistance develops over time as a result of persistently high insulin levels. If fiber can protect against elevated insulin, then it should protect against type 2 diabetes, right? That's *exactly* what the studies show.[22]

The Nurses' Health Studies I and II monitored the dietary records of thousands of women over many decades, and confirmed the protective effect of cereal-fiber intake.[23,24] Women who ate a high-glycemic index diet but also ate large amounts of cereal fiber are protected against type 2 diabetes. In essence, this diet is simultaneously high in 'poison' and in 'antidote.' The two cancel each other out with no net effect. Women who ate a low-glycemic index diet (low 'poison') but also a low-fiber diet (low 'antidote') were also protected. Again the two cancel each other out.

But the deadly combination of a high-glycemic index diet (high 'poison') with low fiber (low 'antidote') increased the risk of type 2 diabetes by a horrifying 75 per cent. *This combination mirrors the exact effect of processing carbohydrates:* processing increases their glycemic index but decreases their fiber content.

The massive 1997 Health Professionals Follow-up Study followed 42,759 men over six years, with essentially the same results.[25] The diet high in glycemic load ('poison') and low in fiber ('antidote') increases the risk of type 2 diabetes by 217 per cent.

The Black Women's Health Study demonstrated that a high-glycemic index diet was associated with a 23 per cent increased risk of type 2 diabetes. A high fiber intake, by contrast, was associated with an 18 per cent *lower* risk of diabetes.

Carbohydrates in their natural, whole, unprocessed form, perhaps with the exception of honey, always contain fiber—which is precisely why junk food and fast food are so harmful. The processing and the addition of chemicals change the food into a form that our bodies have not evolved to handle. That is exactly why these foods are toxic.

One other traditional food may help protect against the modern evils of elevated insulin: vinegar.

THE WONDERS OF VINEGAR

THE WORD VINEGAR originates from the Latin words vinum acer, meaning sour wine. Wine, left undisturbed, eventually turns into vinegar (acetic acid). Ancient peoples quickly discovered vinegar's versatility. Vinegar is still in widespread use as a cleaning substance. Traditional healers exploited the antimicrobial properties of vinegar in a time before antibiotics by using it to clean wounds. Unfiltered vinegar contains 'mother,' which consists of the protein, enzymes and bacteria used to make it.

Vinegar has long been used to preserve food by pickling. As a beverage, the tangy, sour taste of vinegar never gained much popularity, although Cleopatra was famously rumored to drink vinegar in which pearls had been dissolved. However, vinegar still retains fans as a condiment for French fries, a component in dressings (balsamic vinegar) and in making sushi rice (rice vinegar).

Diluted vinegar is a traditional tonic for weight loss. Mention of this folk remedy is found as far back as 1825. British poet Lord Byron popularized vinegar as a weight-loss tonic and would reportedly go for days eating biscuits and potatoes soaked in vinegar.[26] Other ways to use vinegar are to ingest several teaspoons of it prior to meals, or to drink it diluted in water at bedtime. Apple cider vinegar seems to have gained a particular following, as it contains both vinegar (acetic acid) as well as the pectins from the apple cider (a type of soluble fiber).

There are no long-term data on the use of vinegar for weight loss. However, smaller short-term human studies suggest that vinegar may help reduce insulin resistance.[27] Two teaspoons of vinegar taken with a high-carbohydrate meal lowers blood sugar and insulin by as much as 34 per cent, and taking it just before the meal was more effective than taking it five hours before meals.[28] The addition of vinegar for sushi rice lowered the glycemic index of white rice by almost 40 per cent.[29] Addition of pickled vegetables and fermented soybeans (nattō) also significantly lowered the glycemic index of the rice. In a similar manner, rice with the substitution of pickled cucumber for fresh showed a decrease in its glycemic index by 35 per cent.[30]

Potatoes, served cold and dressed with vinegar as a salad, showed considerably lower glycemic index than regular potatoes. The cold storage may favor the development of resistant starch, and the vinegar adds to the benefits. Both glycemic and insulin index were reduced by 43 per cent and 31 per cent respectively.[31] The total amount of carbohydrate is the same in all cases. Vinegar does not displace the carbohydrate, but actually seems to exert a protective effect on the serum insulin response.

Type 2 diabetics drinking two tablespoons of apple cider vinegar diluted in water at bedtime reduced their fasting morning blood sugars.[32] Higher doses of vinegar also seem to increase satiety, resulting in slightly lower caloric intake through the rest of the day (approximately 200 to 275 calories less). This effect was also noted for peanut products. Interestingly, peanuts also resulted in a reduction of glycemic response by 55 per cent.[33]

It's not known how acetic acid produces these beneficial effects. The acid may interfere with the digestion of starches by inhibiting salivary amylase. Vinegar may also reduce the speed of gastric emptying. The data is conflicting, with at least one study showing a 31 per cent reduction in glucose response with no significant delayed gastric emptying.[34]

The use of oil and vinegar dressing is associated with lower risk of cardiovascular disease. The benefit was originally attributed to the effect of dietary alpha linolenic acid. However, Dr. F. Hu of Harvard University points out that mayonnaise, containing similar amounts of alpha linolenic acid, does not appear to provide nearly the same cardiac protection.[35] Perhaps the difference here is the consumption of vinegar. While far from conclusive, it is certainly an interesting hypothesis. Just don't expect rapid weight loss with the use of vinegar. Even its proponents claim only a mild decrease in weight.

THE PROBLEM WITH THE GLYCEMIC INDEX

THE CLASSIFICATION OF carbohydrate foods into the glycemic index was logical and successful. Designed originally for diabetic patients, the index helped them make food choices. However, for the treatment of obesity, low-glycemic index diets have met with mixed success. Weight-reduction benefits have been elusive. And that's because there is one particularly insurmountable problem with the glycemic index diet.

Blood glucose does not drive weight gain. But hormones—particularly insulin and cortisol—do.

Insulin causes obesity. The goal should therefore be to lower *insulin* levels—not glucose levels. The unspoken assumption is that glucose is the only stimulant to insulin secretion. This turns out not to be true at all. There are many factors that raise and lower insulin, especially protein.

(17)

PROTEIN

•

IN THE MID 1990s, as popular sentiment began to turn against the poor, unloved carbohydrate, a backlash originated within the medical community. 'Carbohydrate-reduced diets are nutritionally unbalanced,' they sputtered. That sure sounded good. There are only three macronutrients, after all: protein, fat and carbohydrate. Severe restriction of any one of these runs the risk of an 'unbalanced' diet. Of course, nutritional authorities had no similar compunction about severely restricting dietary fat. But that's beside the point. Certainly any such diet is unbalanced. The more important concern is whether such diets are unhealthy.

So for the sake of argument, let's say that carb-reduced diets are unbalanced. Does that imply that the nutrients contained within carbohydrates are essential for human health?

Certain nutrients are considered essential in our diet because our bodies cannot synthesize them. Without dietary sources of these nutrients, we suffer ill health. There are essential fatty acids, such as the omega 3 and omega 6 fats, and essential amino acids, such as phenylalanine, valine and threonine. But there are no essential carbohydrates and no essential sugars. Those are not required for survival.

Carbohydrates are just long chains of sugars. There is nothing intrinsically nutritious about them. Low-carbohydrate diets that focus on removing refined grains and sugars should be inherently healthier. Perhaps unbalanced, but not unhealthy.

Another criticism leveled at the low-carb diets is that much of the initial weight loss that dieters experience is water—which is true. High carbohydrate intake increases insulin, and insulin stimulates the kidney to reabsorb water. Lowering insulin therefore causes excretion of the excess water. But why is this bad? Who wants swollen ankles?

By the late 1990s, as the 'new' low-carbohydrate approach fused with the prevailing low-fat religion, the Atkins diet v2.0 was born—a low-carb, low-fat and high-protein approach. Where the original Atkins diet was high in fat, this new bastard diet was high in protein. Most high-protein foods also tend to high fat too. But this new approach called for lots of boneless, skinless chicken breasts and egg-white omelets. Once you tired of that, there were protein bars and shakes. A high-protein diet made many worry about potential kidney damage.

High-protein diets are not recommended for those with chronic kidney disease, since the ability to deal with the breakdown products of proteins is impaired. However, in people with normal kidney function, there are no concerns. Several recent studies have concluded that a high-protein diet was not associated with any noticeable harmful effects on kidney function.[1] The concerns about kidney damage were overblown.

The biggest problem with high-protein diets was that they *didn't really work for weight loss.* But why not? The reasoning seems solid. Insulin causes weight gain. Reducing refined carbohydrates lowers blood sugar and insulin. *But all foods cause insulin secretion.* The Atkins v2.0 approach assumed that dietary proteins do not raise insulin since they do not raise blood sugars. This notion was incorrect.

The insulin response to specific foods can be measured and ranked. The glycemic index measures the rise in blood sugar in response to a standard portion of food. The insulin index, created by Susanne Holt in 1997, measures the rise in *insulin* in response to a standard portion of food, and it turns out to be quite different from the glycemic index.[2] Not surprisingly, refined carbohydrates cause a surge in insulin levels. What was astounding was that dietary proteins could cause a similar surge. The glycemic index does not consider protein or fats at all because they do not raise glucose, and that approach essentially ignores the fattening effects of two out of the three major macronutrients. *Insulin can increase independently of blood sugar.*

With carbohydrates, there is a very tight correlation between blood glucose and insulin levels. But overall, blood glucose was responsible for only 23 *per cent* of the variability in the insulin response. The vast majority of the insulin response (77 per cent) has nothing to do with blood sugars. Insulin, not glucose, drives weight gain, and that changes everything.

This point is precisely where glycemic index diets failed. They targeted the glucose response with the assumption that insulin mirrored glucose. But this is not the case. You could lower the *glucose* response, but you didn't necessarily lower the *insulin* response. In the end, the insulin response is what matters.

What factors (other than glucose) determine the insulin response? Consider the incretin effect and cephalic phase.

THE INCRETIN EFFECT AND THE CEPHALIC PHASE

BLOOD SUGAR IS often assumed to be the only stimulus for insulin secretion. But we've long known this was false. In 1966, studies showed that oral or intravenous administration of the amino acid leucine causes insulin secretion.[3] This inconvenient fact was promptly forgotten until it was rediscovered decades later.[4]

In 1986, Dr. Michael Nauck noticed something very unusual.[5] A subject's blood sugar response is identical whether a dose of glucose is given by mouth or intravenously. But, despite the same level of blood sugar, the subject's *insulin* levels differ greatly. Remarkably, the insulin response to oral glucose was much more powerful.

Oral administration almost never has a stronger effect than intravenous. Intravenous infusions have 100 per cent bioavailability, meaning that all of the infusion is delivered directly into the blood. When given by mouth, many medicines are incompletely absorbed or partially deactivated by the liver before reaching the bloodstream. For this reason, intravenous delivery tends to be much more effective.

However, in this situation, the opposite was true. Oral glucose was far, far better at stimulating insulin. Furthermore, *this mechanism had nothing to do with the blood sugar level.* This phenomenon had not previously been described. Intensive research efforts revealed that the stomach itself produces hormones called incretins that increase insulin secretion. Since the intravenous glucose bypasses the stomach, there is no incretin effect. The incretin effect may account for *50 per cent to 70 per cent* of the insulin secretion after oral glucose intake.

Rather than simply being a mechanism for food absorption and excretion, the gastrointestinal tract, with its nerve cells, receptors and hormones, functions almost like a 'second brain.' The two human incretin hormones described so far are glucagon-like peptide 1 (GLP-1) and glucose-dependent insulinotropic polypeptide (GIP). Both hormones are deactivated by the hormone dipeptidyl peptidase-4. The incretins are secreted by the stomach and small intestine in response to food. Both GLP-1 and GIP increase insulin release by the pancreas. Fats, amino acids and glucose *all* stimulate incretin release and thus, increase insulin levels. Even non-nutritive sweeteners, which have no calories at all, can stimulate the insulin response. Sucralose in humans, for example, raises the insulin level 22 per cent higher.[6]

The incretin effect starts within minutes of ingestion of nutrients into the stomach and peaks at roughly sixty minutes. The incretins have other important effects as well. They delay emptying of stomach contents into the small intestine, which slows down glucose absorption.

The cephalic phase is another pathway of insulin secretion independent of glucose. The body anticipates food as soon as it goes in your mouth and long before nutrients hit the stomach. For example, swishing a sucrose or saccharin solution around your mouth and spitting it out will increase your insulin level.[7] While the importance of the cephalic phase is unknown, it highlights the significant fact that there are multiple glucose-independent pathways of insulin release.

The discovery of these new pathways was electrifying. The incretin effect explains how fatty acids and amino acids also play a role in stimulating insulin. All foods, not just carbohydrates, stimulate insulin. Thus, *all foods can cause weight gain.* And hence we get major confusion with calories. High-protein foods can cause weight gain—not due to their caloric content, but rather to their insulin-stimulating effects. If carbohydrates are not the only or even the major stimulus to insulin, then restricting carbohydrates may not always be as beneficial as we believed. Substituting insulin-stimulating proteins for insulin-stimulating carbohydrates produces no net benefit. Dietary fat, though, tends to have the weakest insulin-stimulating effect.

DAIRY, MEAT AND THE INSULIN INDEX

PROTEINS DIFFER GREATLY in their capacity to stimulate insulin,[8] with dairy products in particular being potent stimuli.[9] Dairy also shows the largest discrepancy between the blood glucose and insulin effect. It scores extremely low on the glycemic index (15 to 30), but very high on the insulin index (90 to 98). Milk does contain sugars, predominantly in the form of lactose. However, when tested, pure lactose has minimal effect on either the glycemic or insulin indexes.

Milk contains two main types of dairy protein: casein (80 per cent) and whey (20 per cent). Cheese contains mostly casein. Whey is the byproduct left over from the curds in cheese making. Bodybuilders frequently use whey protein supplements because it is high in branched-chain amino acids, felt to be important in muscle formation. Dairy protein, particularly whey, is responsible for raising insulin levels even higher than whole-wheat bread, due largely to the incretin effect.[10] Whey protein supplementation increased GLP-1 by 298 per cent.[11]

The insulin index shows great variability, but nevertheless, there are some general patterns. Increasing carbohydrate consumption leads to increased insulin secretion. This relationship forms the basis of many low-carbohydrate and glycemic index diets, and also explains the well-known propensity of starchy and sugary foods to cause obesity.

Fatty foods can also stimulate insulin, but pure fats, such as olive oil do not stimulate insulin or glucose. However, few foods are eaten as pure fat. It may be that the protein component of fatty foods drives the insulin response. It is also interesting that fat tends to have a flat dose-response curve. Higher and higher amounts of fat do not stimulate any greater insulin response. Despite the higher caloric value of fat, it stimulates insulin less than carbohydrates or protein.

The surprise here is dietary protein. The insulin response is highly variable. While vegetable proteins raise insulin minimally, whey protein and meat (including seafood) cause significant insulin secretion. But are dairy and meat are fattening? That question is complicated. The incretin hormones have multiple effects, only *one* of which is to stimulate insulin. Incretins also have a major effect on satiety.

SATIETY

INCRETIN HORMONES PLAY an important role in the control of gastric emptying. The stomach normally holds food and mixes it with stomach acid before slowly discharging the contents. GLP-1 causes stomach emptying to significantly slow. Absorption of nutrients also slows, resulting in lower blood glucose and insulin levels. Furthermore, this effect creates a sensation of satiety that we experience as 'being full.'

A 2010 study compared the effect of four different proteins: eggs, turkey, tuna and whey protein—on participants' insulin levels.[12] As expected, whey resulted in the highest insulin levels. Four hours afterward, participants were treated to a buffet lunch. The whey group ate substantially less than the other groups. The whey protein suppressed their appetites and increased their satiety. In other words, those subjects were 'full.' See Figure 17.1.[13]

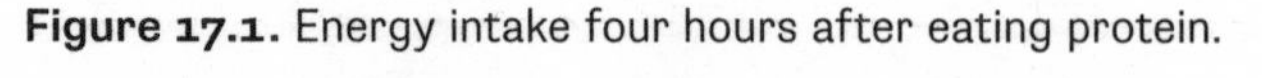

Figure 17.1. Energy intake four hours after eating protein.

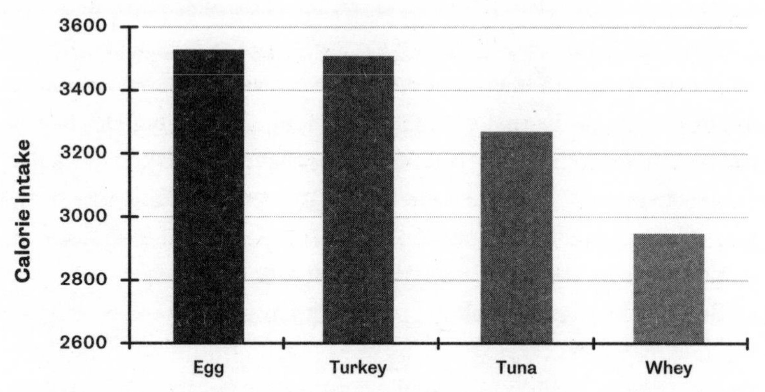

So the incretin hormones produce two opposing effects. Increased insulin promotes weight gain, but increased satiety suppresses it—which is consistent with personal experience. Animal proteins tend to cause you to feel fuller for longer, with whey having the greatest effect. Compare two calorically equal portions of food: a small steak versus a large sugared soda. Which keeps you full longer? The clear winner is the steak. It creates more satiety. The steak just 'sits' in your stomach. You are feeling the incretin effect of slowing the emptying of stomach contents. The soda, however, does not 'sit' in your stomach for long, and you quickly become hungry again.

These two opposing effects—insulin promotes weight gain, but satiety promotes weight loss—cause a maddening debate about meat and dairy. The important question is this: Which effect is more powerful? It is possible that the specific incretin stimulated may be important in determining weight gain or loss. For example, selective stimulation of GLP-1, as with a drug such as exenatide, produces weight loss, as the resulting satiety effects outweigh the weight-gaining effects.

Therefore, we must consider each protein separately since there is considerable variation in the effect of each on weight. The main dietary proteins studied are dairy and meat, and we have two main considerations here: the incretin effect and the portion of dietary protein.

MEAT

TRADITIONALLY, MEAT CONSUMPTION was thought to cause weight gain because meat is high in protein, fat and calories.[14] However, more recently, many consider it to cause weight *loss* because it is low in carbohydrate. Which is true? This is a difficult question since the only data available are association studies, which are open to interpretation and cannot establish causation.

The European Prospective Investigation into Cancer and Nutrition study was a massive European prospective cohort study started in 1992, encompassing 521,448 volunteers from ten countries. After five years of follow-up, findings showed that total meat, red meat, poultry and processed meats were all significantly associated with weight gain, even after adjustment for total caloric intake.[15,16] Eating three extra servings of meat per day is associated with an extra pound in weight gain over one year, even after controlling for calories.

In North America, combined data is available from the Nurses' Health Studies I and II and the Health Professionals Follow-up Study.[17] Both processed and unprocessed red meat were associated with weight gain. Each additional daily serving of meat increased body weight by approximately 1 pound (0.45 kilograms). This effect even exceeded the weight-increasing effect of sweets and desserts! So, on balance, the weight-increasing effect seems predominant here. There are some possible contributing factors.

First, most beef is now raised in a feedlot and fed grain. Cows are ruminants that naturally eat grass. This change in their diet may change the character of the meat.[18] Wild-animal meat is similar to grass-fed beef, but not to grain-fed beef. Feedlot cattle require large doses of antibiotics. Farm-raised fish also have little in common with wild fish. Farm-raised fish eat pellets that often contain grains and other cheap substitutes for a fish's natural diet.

Second, while we understand the benefits of eating 'whole' foods, we do not apply this knowledge to meat. We eat only the muscle meats rather than the entire animal, thereby risking overconsumption of the muscle meats. We generally discard most of the organ meats, cartilage and bones—which is analogous to drinking the juice of a fruit but discarding the pulp. Yet bone broth, liver, kidney and blood are all parts of the traditional human diets. Traditional staples like steak-and-kidney pie, blood sausage and liver have disappeared. Ethnic foods such as tripe, pork bung, congealed pig's blood, oxtail and beef tongue still survive.

The organ meats tend to be the fattiest parts of the animal. By focusing almost exclusively on the muscles of animals for food, we are preferentially eating protein rather than fat.

DAIRY

THE STORY WITH dairy is entirely different. Despite the fact that its consumption causes big increases in insulin levels, large observation studies do *not* link dairy to weight gain. If anything, dairy protects against weight gain, as found in the Swedish Mammography Cohort.[19] In particular, whole milk, sour milk, cheese and butter were associated with less weight gain, but not low-fat milk. The ten-year prospective CARDIA Study found that the highest intake of dairy is associated with the lowest incidence of obesity and type 2 diabetes.[20] Other large population studies confirmed this association.[21,22]

The data from the Nurses' Health Studies and the Health Professionals Follow-up Study[23] shows that overall, average weight gain over any four-year period was 3.35 pounds (1.5 kilograms)—pretty close to 1 pound per year. Milk and cheese were essentially weight neutral. Yogurt seemed to be particularly slimming, possibly due to the fermentation process. Butter did have a small effect on weight gain.

Why is there such a difference between dairy and meat? One difference is portion size. Eating more meat is easy. You could eat a large steak or half a roast chicken or a large bowl of chili. However, increasing dairy protein to the same degree is more difficult. Can you eat a huge slab of cheese for dinner? How about drink several gallons of milk? Eat two large tubs of yogurt for lunch? Hardly. It is difficult to significantly increase dairy proteins without resorting to whey protein shakes and other such artificial foods. An extra glass of milk a day doesn't cut it. Therefore, even if dairy proteins are particularly good at stimulating insulin, the small portions do not make a large overall difference.

By eating large amounts of skim milk, lean meat and protein bars, Atkins enthusiasts were unintentionally stimulating their insulin to the same degree as before. Substituting large amounts of lean, often processed meat for carbohydrates was not a winning strategy.[24] Reducing sugar and white bread was good advice. But replacing them with luncheon meats was not. Furthermore, with increased meal frequency, the protection of the incretin effect was diminished.

THE HORMONAL OBESITY THEORY

NOW WE CAN modify the hormonal obesity theory to include the incretin effect to provide the complete picture, as illustrated in Figure 17.2.

Figure 17.2. The hormonal obesity theory.

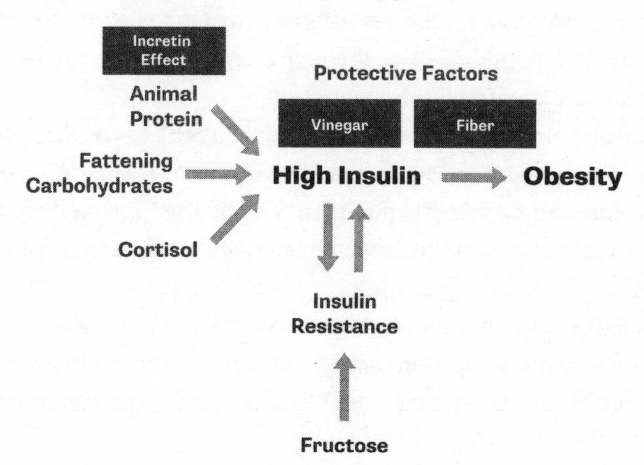

Animal protein is highly variable but comes with the protective effect of satiety. And we shouldn't ignore the protective power of the incretin effect. The slowing of gastric motility increases satiety so that we feel more full and therefore eat less at the next meal, or even skip a meal altogether to allow ourselves 'time to digest.' This behavior is instinctive. When children are not hungry, they will not eat. Wild animals also show the same restraint. But we've trained ourselves to ignore our own feelings of satiety so that we will eat when the time comes, whether we are hungry or not.

Here's a small tip for weight loss, one that should be obvious, but is not. If you are not hungry, don't eat. Your body is telling you that you should not be eating. After indulging in a large meal, such as we do at Thanksgiving, we feel paranoid about skipping the next meal because of an irrational fear that missing even a single meal will wreck our metabolism. So we circumvent the protective effect of incretins by rigidly scheduling three meals a day with snacks, come hell or high water.

There is still more to learn. Blood glucose accounts for only 23 per cent of the insulin response. Dietary fats and protein only accounts for another 10 per cent. Close to *67 per cent* of the insulin response is still unknown—which is tantalizingly close to the 70 per cent contribution to obesity that is inherited, as described in chapter 2. Other suspected factors include presence of dietary fiber, an elevated amylose/amylopectin ratio, preserved botanical integrity (whole foods), presence of organic acids (fermentation), addition of vinegar (acetic acid) and addition of chili peppers (capsaicin).

Simplistic arguments that 'Carbs make you fat!' or 'Calories make you fat!' or 'Red meat makes you fat!' or 'Sugar makes you fat!' do not fully capture the complexity of human obesity. The hormonal obesity theory provides a framework for understanding the interaction of disease.

All foods stimulate insulin, thus all foods could be fattening—*and that's where the calorie confusion emerges.* Since all foods could be fattening, we imagined that all foods could be measured in a common unit:

the calorie. But the calorie was the wrong unit. Calories do not cause obesity. Instead, insulin is responsible. Without a framework for understanding insulin, it was impossible to understand the inconsistencies of the epidemiologic evidence. The low-fat caloric-reduction approach was a proven failure. The high-protein approach was subsequently proved a failure. And so many returned to the failed caloric-reduction approach.

But a new approach known as the Paleo diet—sometimes referred to as the 'caveman diet' or the 'original human diet'—was gaining strength. Only foods that were available in Paleolithic or ancient times are to be consumed. Dieters avoid all processed foods, added sugars, dairy, grains, vegetable oils, sweeteners and alcohols. However, fruits, vegetables, nuts, seeds, spices, herbs, meats, seafood and eggs are all acceptable. The Paleo diet does not limit carbohydrates, proteins or fat. Instead, the consumption of *processed* foods is curtailed. Remember that the single defining characteristic of the Western diet is food processing, not macronutrient content. *The toxicity lies not in the food, but in the processing.*

The Low Carb, High Fat or Low Carb, Healthy Fat (LCHF) diet is similar, maintaining a focus on real foods. The main difference is that the LCHF diet allows dairy products and is stricter on the fruits due to their carbohydrate content. The LCHF approach makes some sense, as dairy foods are generally not associated with weight gain. This factor allows for greater dietary choice and, hopefully, better long-term compliance.

The Paleo/LCHF diet is based on the simple observation that humans can eat a wide variety of foods without becoming obese or developing diabetes. These foods can be eaten without counting calories, counting carbohydrates or using food diaries, pedometers or any other such artificial means. You simply eat when hungry and don't eat when full. However, the foods are all of an unprocessed nature and had been consumed by humans for thousands of years without causing illness. They had withstood the test of time. These are the foods that we should base our diets on.

There are no intrinsically bad foods, only processed ones. The further you stray from real food, the more danger you are in. Should you eat protein bars? No. Should you eat meal replacements? No. Should you drink meal replacement shakes? Absolutely not. Should you eat processed meats, processed fats or processed carbohydrates? No, no and no.

While we ideally would all eat grass-fed beef and organically raised strawberries, let's be real here. There will be times when we'll eat processed food because it is cheap, available and, let's face it, delicious (think ice cream). However, we have, over the centuries, developed other dietary strategies such as fasting to detoxify or cleanse ourselves. These strategies have been lost in the mists of time. We will rediscover these ancient secrets soon, but for now, stick to real food.

Natural foods contain significant amounts of saturated fats. This fact naturally leads to the questions: Won't all these saturated fats clog up my arteries? Won't it lead to heart attacks? The short answer is 'No.'

But why not? That's the subject of the next chapter.

(18)

FAT PHOBIA

It is now increasingly recognized that the low-fat campaign has been based on little scientific evidence and may have caused unintended health consequences.

HARVARD RESEARCHERS DRS. FRANK HU & WALTER WILLETT, 2001

ONE OF THE giants of modern nutritional science, Dr. Ancel Keys (1904–2004) received his first PhD in oceanography and biology and a second PhD in physiology from the University of Cambridge. He spent most of the rest of his career at the University of Minnesota, where he would play a dominant role in defining the current nutritional landscape.

During World War II, Dr. Keys led the development of the K-rations, which would form the basis of military nutrition in the United States. He studied the effects of severe caloric restriction in the famous Minnesota Starvation Experiment (discussed in chapter 3). However, his crowning achievement is considered to be the Seven Countries Study, a long-term observational study of diet and heart disease.

In the post–World War II years, starvation and malnutrition were the major nutritional challenges. But Dr. Keys was struck by an odd inconsistency. Americans, despite far better nourishment, were suffering from rising rates of heart attack and stroke. In war-ravaged Europe, those rates remained low.[1] In 1951, Dr. Keys noticed low rates of heart disease in Italian laborers. The Mediterranean diet, as he observed in Naples, was substantially lower in fat (20 per cent of calories) than the American diet of the period (approximately 45 per cent of calories).[2] Most striking, though, was the lower rate of consumption of animal foods and saturated fat. He hypothesized that high blood cholesterol levels caused heart disease and that the low dietary intake of fat was protective. In 1959, he published his dietary advice for the prevention of cardiovascular disease.[3] Prominent among his recommendations were the following guidelines:

- Do not get fat; if you are fat, reduce. (Easier said than done!)
- Restrict saturated fats; the fats in beef, pork, lamb, sausages, margarine and solid shortenings; and the fats in dairy products.
- Prefer vegetable oils to solid fats, but keep total fats under 30 per cent of your diet calories.

These recommendations survived relatively intact and defined nutritional orthodoxy for the next half century. In 1977, they became enshrined in the *Dietary Guidelines for Americans*.[4] The main message, then as now, is that all fat is bad, but saturated fats particularly so. Dietary fat was thought to 'clog up the arteries' and cause heart attacks.

The ambitious Seven Countries Study compared rates of coronary disease with various diet and lifestyle factors across nations. By 1970, with five years' worth of data, the study had several main conclusions regarding fats:[5]

- Cholesterol levels predicted heart-disease risk.
- The amount of saturated fat in the diet predicted cholesterol levels.
- Monounsaturated fat protected against heart disease.
- The Mediterranean diet protected against heart disease.

Significantly, total dietary fat was *not* correlated to heart disease. Rather, saturated fat was dangerous, but mono-unsaturated fats were protective. Dietary cholesterol was also *not* identified as a risk factor for heart disease.

Heart disease is caused by atherosclerosis—the process by which arteries in the heart become narrowed and hardened by the buildup of plaque. But atherosclerosis is not simply the result of high cholesterol levels clogging arteries. Current opinion holds that plaque develops as a response to injury: the wall of the artery becomes damaged, resulting in inflammation, which in turn allows infiltration of cholesterol and inflammatory cells into artery walls, in addition to the proliferation of smooth muscle. The narrowing of the artery may cause chest pain (also called angina). When plaques rupture, a blood clot forms, which abruptly blocks the artery. The resulting lack of oxygen causes a heart attack. Heart attacks and strokes are predominantly inflammatory diseases, rather than simply diseases of high cholesterol levels.

This understanding, however, came much later. In the 1950s, it was imagined that cholesterol circulated and deposited on the arteries much like sludge in a pipe (hence the popular image of dietary fat clogging up the arteries). It was believed that eating saturated fats caused high cholesterol levels, and high cholesterol levels caused heart attacks. This series of conjectures became known as the diet-heart hypothesis. Diets high in saturated fats caused high blood cholesterol levels, which caused heart disease.

The liver manufactures the overwhelming majority—80 per cent—of the blood cholesterol, with only 20 per cent coming from diet. Cholesterol is often portrayed as some harmful poisonous substance that must be eliminated, but nothing could be farther from the truth. Cholesterol is a key building block in the membranes that surround all the cells in our body. In fact, it's so vital that every cell in the body except the brain has the ability to make it. If you reduce cholesterol in your diet, your body will simply make more.

The Seven Countries Study had two major problems, although neither was very obvious at the time. First, it was a correlation study. As such, its findings could not prove causation. Correlation studies are dangerous because it is very easy to mistakenly draw causal conclusions. However, they are often the only source of long-term data available. It is always important to remember that they can only generate hypotheses to be tested in more rigorous trials. The heart benefit of the low-fat diet was not proven false until 2006 with the publication of the Women's Health Initiative Dietary Modification Trial and the Low-Fat Dietary Pattern and Risk of Cardiovascular Disease study,[6] some thirty years after the low-fat approach became enshrined in nutritional lore. By that time, like a supertanker, the low-fat movement had gained so much momentum that it was impossible to turn it aside.

The *association* of heart disease and saturated fat intake is not proof that saturated fat *causes* heart disease. Some recognized this fatal flaw immediately[7] and argued against making dramatic dietary recommendations based on such flimsy evidence. The seemingly strong link between heart disease and saturated fat consumption was forged with quotation and repetition, not with scientifically sound evidence. There were many possible interpretations of the Seven Countries Study. Animal protein, saturated fats and sugar were all correlated to heart disease. Higher sucrose intake could just as easily have explained the correlation to heart disease, as Dr. Keys himself had acknowledged.

It is also possible that higher intakes of animal protein, saturated fats and sugar are all merely markers of industrialization. Counties with higher levels of industrialization tended to eat more animal products (meat and dairy) and also tended to have higher rates of heart disease. Perhaps it was the processed foods. All of these hypotheses could have been generated from the same data. But what we got was the diet-heart hypothesis and the resulting low-fat crusade.

The second major problem was the inadvertent triumph of nutritionism, a term popularized by the journalist and author Michael Pollan.[8] Rather than discussing individual foods (spinach, beef, ice cream), nutritionism reduced foods to only three macronutrients: carbohydrates, proteins and fats. They were then subdivided further as saturated and unsaturated fats, trans fats, simple and complex carbohydrates, etc. This sort of simplistic analysis does not capture the hundreds of nutrients and phytochemicals in foods, all of which affect our metabolism. Nutritionism ignores the complexity of food science and human biology.

An avocado, for instance, is not simply 88 per cent fat, 16 per cent carbohydrate and 5 per cent protein with 4.9 grams of fiber. This sort of nutritional reductionism is how avocados became classified for decades as a 'bad' food due to their high fat content, only to the reclassified today as a 'super food.' Nutritionally, a piece of butterscotch candy cannot be reasonably compared to kale simply because both contain equal amounts of carbohydrate. Nutritionally, a teaspoon of trans-fat–laden margarine cannot be reasonably compared to an avocado simply because both contain equal amounts of fats.

Dr. Keys made the unnoticed and unintentional claim that all saturated fats, all unsaturated fats, all dietary cholesterol, etc., are the same. This fundamental error led to decades of flawed research and understanding. Nutritionism fails to consider foods as individuals, each with its own particular good and bad traits. Kale is not the same nutritionally as white bread, even though both contain carbohydrates.

These two fundamental but subtle errors of judgment led to the widespread acceptance of the diet-heart hypothesis, even though the evidence supporting it was shaky at best. Most natural animal fats are chiefly composed of saturated fats. In contrast, vegetable oils such as corn are chiefly omega 6 polyunsaturated fatty acids.

After remaining relatively stable from 1900 to 1950, animal-fat consumption began a relentless decline. The dialogue began to change in the late 1990s due to popularity of higher-fat diets. The unintended consequence of the saturated fat reduction was that intake of omega 6 fatty acids increased significantly. Carbohydrates, as a percentage of calories, also started to climb. (To be more precise, these were *intended* consequences. They were unintentionally detrimental to human health.)

Omega 6s are a family of polyunsaturated fatty acids that are converted to highly inflammatory mediators called eicosanoids. The massive increase in the use of vegetable oils can be traced to technological advances in the 1900s that allowed modern production methods. Since corn is not naturally high in oil, normal human consumption of omega 6 oils had been quite low. But now we could process literally tons of corn in order to derive useful amounts.

Omega 3 fatty acids are another family of polyunsaturated fats that are mainly anti-inflammatory. Flax seeds, walnuts and oily fish such as sardines and salmon are all good sources. Omega 3 fatty acids decrease thrombosis (blood clots) and are believed to protect against heart disease. Low rates of heart disease were originally described in the Inuit population and subsequently in all major fish-eating populations.

High dietary ratios of omega 6:3 ratios increase inflammation, potentially worsening cardiovascular disease. It is estimated that humans evolved eating a diet that is close to equal in omega 6 and 3 fatty acids.[9] However, the current ratio in the Western diet is closer to a 15:1 to 30:1 ratio. Either we are eating way too little omega 3, way too much omega 6, or more likely, both. In 1990, the Canadian nutritional guidelines were the first to recognize the important difference and include specific recommendations for both types of fatty acids. Animal fats had been replaced by highly inflammatory omega 6–laden vegetable oils that had been widely advertised as 'heart healthy.' This is ironic since atherosclerosis is now considered mostly to be an inflammatory disease.

To replace butter, Americans increasingly reached for that tub of edible plastic: margarine. With large advertising campaigns designed to play up its wholesome all-vegetable origins, trans-fat–laden margarine was in the right place at the right time. Designed in 1869 as a cheap butter alternative, it was originally made from beef tallow and skim milk. Margarine is naturally an unappetizing white, but is dyed yellow. Butter manufacturers were not amused, and marginalized margarine for decades through tariffs and laws. Its big break came with World War II and the ensuing butter shortage. Most of the taxes and laws against margarine were repealed since butter was scarcely available anyway.

This action paved the way for the great margarine renaissance of the 1960s and '70s as the war on saturated fats gained ground. Ironically, this 'healthier' alternative, chock full of trans fats, was actually killing people. Thankfully, consumer advocacy forced the retreat of trans fats from store shelves.

It's actually a minor miracle that vegetable oils were considered healthy at all. Squeezing oil from non-oily vegetables requires a substantial amount of industrial-strength processing, including pressing, solvent extraction, refining, degumming, bleaching and deodorization. There is nothing natural about margarine and it could only have become popular during an era in which artificial equaled good. We drank artificial orange juices like Tang. We gave our children artificial baby formula. We drank artificially sweetened sodas. We made Jell-O. We thought we were smarter than Mother Nature. Whatever she had made, we could make better. Out with all-natural butter. In with industrially produced, artificially colored trans-fat–laden margarine! Out with natural animal fats. In with solvent-extracted, bleached and deodorized vegetable oil! What could possibly go wrong?

THE DIET-HEART HYPOTHESIS

IN 1948, HARVARD University began a decades-long community-wide prospective study of the diets and habits of the town of Framingham, Massachusetts. Every two years, all residents would undergo screening with blood work and questionnaires. High cholesterol levels in the blood had been associated with heart disease. But what caused this increase? A leading hypothesis was that high dietary fat was a prime factor in raising cholesterol levels. By the early 1960s, the results of the Framingham Diet Study were available. Hoping to find a definitive link between saturated-fat intake, blood cholesterol and heart disease, the study instead found... nothing at all.

There was absolutely no correlation. Saturated fats did not increase blood cholesterol. The study concluded, 'No association between per cent of calories from fat and serum cholesterol level was shown; nor between ratio of plant fat to animal fat intake and serum cholesterol level.'

Did saturated fat intake increase risk of heart disease? In a word, no. Here are the final conclusions of this forgotten jewel: *'There is, in short, no suggestion of any relation between diet and the subsequent development of CHD [coronary heart disease] in the study group.'*[10]

This negative result would be repeatedly confirmed over the next half century. No matter how hard we looked, there was no discernible relationship between dietary fat and blood cholesterol.[11] Some trials, such as the Puerto Rico Heart Health Program, were huge, boasting more than 10,000 patients. Other trials lasted more than twenty years. The results were always the same. Saturated-fat intake could not be linked to heart disease.[12]

But researchers had drunk the Kool-Aid. They believed their hypothesis so completely that they were willing to ignore the results of their own study. For example, in the widely cited Western Electric Study,[13] the authors note that 'the amount of saturated fatty acids in the diet was not significantly associated with the risk of death from CHD.' This lack of association, however, did not dissuade the authors from concluding 'the results support the conclusion that lipid composition of the diet affects serum cholesterol concentration and risk of coronary death.'

All these findings should have buried the diet-heart hypothesis. But no amount of data could dissuade the diehards that dietary fat caused heart disease. Researchers saw what they wanted to see. Instead, researchers saved the hypothesis and buried the results. Despite the massive effort and expense, the Framingham Diet Study was never published in a peer-reviewed journal. Instead, results were tabulated and quietly put away in a dusty corner—which condemned us to fifty years of a low-fat future that included an epidemic of diabetes and obesity.

There was also the confounding issue of the artificial trans fats.

TRANS FATS

SATURATED FATS ARE so named because they are saturated with hydrogen. This makes them chemically stable. The polyunsaturated fats, like most vegetable oils, have 'holes' where the hydrogen is 'missing.' They are less stable chemically, so they have a tendency to go rancid and have a short shelf life. The solution was to create artificial trans fats.

There are natural trans fats. Dairy products contain between 3 per cent to 6 per cent natural trans fats.[14] Beef and lamb contain a little less than 10 per cent. However, these natural trans fats are not believed to be harmful to human health.

In 1902, Wilhelm Normann discovered that you could bubble hydrogen into vegetable oil to saturate it, turning polyunsaturated fat into saturated fat. Food labels often called this partially hydrogenated vegetable oil. Trans fat is less likely to go rancid. Trans fats are semi-solid at room temperature, so they spread easily and have an improved mouth feel. Trans fats were ideal for deep-frying. You can use this stuff over and over without changing it.

Best of all, this stuff was *cheap*. Using leftover soybeans from animal feed, manufacturers could process the heck out it and still get vegetable oil. A little hydrogen, a little chemistry, and boom—trans fats, baby. So what if it killed millions of people from heart disease? That knowledge was years in the future.

Trans fats started to hit their stride by the 1960s, as saturated fats were fingered as the main cause of heart disease. The makers of trans fats were quick to point out that they were processed from polyunsaturated fats—the 'heart healthy' fat. Trans fats retained a healthy veneer, even as they were killing people left and right. Margarine, another completely artificial food, embraced trans fats like a long-lost lover.

Saturated fat consumption—butter, beef and pig fats—steadily decreased. McDonald's switched from frying in 'unhealthy' beef tallow to frying in trans-fat–laden vegetable oils. Theaters switched from frying in naturally saturated coconut oil to artificially saturated trans fats. Other major sources of trans fats included deep-fried and frozen foods, packaged bakery products, crackers, vegetable shortening and margarine.

The year 1990 marked the beginning of the end for trans fat when Dutch researchers noted that consuming trans fats increased LDL (low-density lipoprotein or 'bad' cholesterol) and lowered HDL (high-density lipoprotein or 'good' cholesterol) in subjects.[15] Closer scrutiny of the health effects led to an estimate that a 2 per cent increase in trans-fat consumption would increase risk of heart disease by a whopping 23 per cent.[16] By 2000, the tide had turned decisively. Most consumers were actively avoiding trans fats, and Denmark, Switzerland and Iceland banned trans fats for human consumption.

The recognition of the dangers of trans fats led to a re-evaluation of previous studies of saturated fats. Previous studies had classified trans fats together with saturated fats. Researchers strove to separate out the effects of trans fats, and that changed everything we thought we knew about saturated fats.

PROTECTIVE EFFECT ON HEART DISEASE AND STROKE

ONCE THE SKEWING effect of trans fats was taken into account, the studies consistently showed that high dietary fat intake was not harmful.[17] The enormous Nurses' Health Study followed 80,082 nurses over fourteen years. After removing the effect of trans fats, this study concluded that 'total fat intake was not significantly related to the risk of coronary disease.'[18] Dietary cholesterol was also safe. The Swedish Malmo Diet and Cancer Study[19] and a 2014 meta-analysis published in the *Annals of Internal Medicine*[20] reached similar conclusions.

And the good news for saturated fats kept rolling in. Dr. R. Krause published a careful analysis of twenty-one studies covering 347,747 patients and found 'no significant evidence for concluding that dietary saturated fat is associated with an increased risk of CHD.'[21] In fact, there was even a small protective effect on stroke. The *protective* effects of saturated fats were also found in the fourteen-year, 58,543-person Japan Collaborative Cohort Study for Evaluation of Cancer and the ten-year Health Professionals Follow-up Study of 43,757 men.[22,23,24]

Ironically, trans-fat laden margarines had always branded themselves as heart healthy since they were low in saturated fat. Twenty-year follow-up data from the Framingham study revealed that margarine consumption was associated with *more* heart attacks. By contrast, eating more butter was associated with *fewer* heart attacks.[25,26]

A ten-year study in Oahu, Hawaii, found a protective effect of saturated fat on stroke risk.[27] The twenty-year follow-up data from the Framingham study confirmed these benefits.[28] Those eating the most saturated fat had the *least* strokes, but polyunsaturated fats (vegetable oils) were not beneficial. Monounsaturated fats (olive oil) were also protective against stroke, a consistent finding throughout the decades.

DIETARY FAT AND OBESITY

THE EVIDENCE ON a link between dietary fat and obesity is consistent: there is no association whatsoever. The main concern about dietary fats had always been heart disease. Obesity concerns were just 'thrown in' as well.

When dietary fat was declared a villain, cognitive dissonance set in. Dietary carbohydrates could not be good (because they are low in fat) and bad (because they are fattening) at the same time. Without anybody even noticing, it was decided that *carbohydrates* were no longer fattening; *calories* were fattening. Dietary fat, with it high caloric density, must therefore be bad for weight gain as well. However, there was never any data to support this assumption.

Even the National Cholesterol Education Program admits, 'The percentage of total fat in the diet, independent of caloric intake, has not been documented to be related to body weight.'[29] Translation: despite fifty years of trying to prove that dietary fat causes obesity, we still cannot find any evidence. This data is hard to find because *it never existed.*

A comprehensive review of all the studies of high-fat dairy finds no association with obesity,[30] with whole milk, sour cream and cheese offering greater benefits than low-fat dairy.[31] Eating fat does not make you fat, but may protect you against it. Eating fat together with other foods tends to decrease glucose and insulin spikes.[32] If anything, dietary fat would be expected to protect against obesity.

While literally thousands of papers have reviewed this data, perhaps Dr. Walter Willett of the Harvard T.H. Chan School of Public Health said it best in his 2002 review article entitled, 'Dietary Fat Plays a Major Role in Obesity: No.'[33] Considered one of the world's foremost experts in nutrition, he writes,

> Diets high in fat do not account for the high prevalence of excess body fat in Western countries; reductions in the percentage of energy from fat will have no important benefits and could further exacerbate this problem. The emphasis on total fat reduction has been a serious distraction in efforts to control obesity and improve health in general.

The failure of the low-fat paradigm was fully exposed in the Women's Health Initiative Dietary Modification Trial.[34] Nearly 50,000 women were randomly assigned to low-fat or regular diets. Over seven years, the low-fat, calorie-restricted diet produced no benefits in weight loss. Nor were there heart-protection benefits either. The incidence of cancer, heart disease or stroke was not reduced. There were no cardiovascular benefits. There were no weight benefits. The low-fat diet was a complete failure. The emperor had no clothes.

PART SIX

The Solution

(19)

WHAT TO EAT

•

THERE ARE TWO prominent findings from all the dietary studies done over the years. First: *all diets work.* Second: *all diets fail.*

What do I mean? Weight loss follows the same basic curve so familiar to dieters. Whether it is the Mediterranean, the Atkins or even the old fashioned low-fat, low-calorie, all diets in the short term seem to produce weight loss. Sure, they differ by amount lost—some a little more, some a little less. But they all seem to work. However, by six to twelve months, weight loss plateaus, followed by a relentless regain, despite continued dietary compliance. In the ten-year Diabetes Prevention Program, for example, there was a 15.4-pound (7-kilogram) weight loss after one year.[1] The dreaded plateau, then weight regain, followed.

So all diets fail. The question is why.

Permanent weight loss is actually a two-step process. There is a short-term and a long-term (or time-dependent) problem. The hypothalamic region of the brain determines the body set weight—the fat thermostat. (For more on body set weight, see chapters 6 and 10.) Insulin acts here to set body set weight higher. In the short term, we can use various diets to bring our actual body weight down. However, once it falls below the body set weight, the body activates mechanisms to regain that weight—and that's the long-term problem.

This resistance to weight loss has been proven both scientifically and empirically.[2] Obese persons that had lost weight required fewer calories because their metabolisms had slowed dramatically and desire to eat accelerates. The body actively resists long-term weight loss.

THE MULTIFACTORIAL NATURE OF DISEASE

THE *MULTIFACTORIAL* NATURE of obesity is the crucial missing link. There is no one single cause of obesity. Do calories cause obesity? Yes, partially. Do carbohydrates cause obesity? Yes, partially. Does fiber protect us from obesity? Yes, partially. Does insulin resistance cause obesity? Yes, partially. Does sugar cause obesity? Yes, partially. (See chapter 17, Figure 17.2.) All these factors converge on several hormonal pathways that lead to weight gain, and insulin is the most important of these. Low-carbohydrate diets reduce insulin. Low-calorie diets restrict all foods and therefore reduce insulin. Paleo and LCHF diets (low in refined and processed foods) reduce insulin. Cabbage-soup diets reduce insulin. Reduced-food-reward diets reduce insulin.

Virtually all diseases of the human body are multifactorial. Consider cardiovascular disease. Family history, age, gender, smoking, high blood pressure and physical activity all influence, perhaps not equally, the development of heart disease. Cancer, stroke, Alzheimer's disease and chronic renal failure are all multifactorial diseases.

Obesity is also a multifactorial disease. What we need is a framework, a structure, a coherent theory to understand how all its factors fit together. Too often, our current model of obesity assumes that there is only one single true cause, and that all others are pretenders to the throne. Endless debates ensue. Too many calories cause obesity. No, too many carbohydrates. No, too much saturated fat. No, too much red meat. No, too much processed foods. No, too much high fat dairy. No, too much wheat. No, too much sugar. No, too much highly palatable foods. No, too much eating out. It goes on and on. They are all partially correct.

So the low-calorie believers disparage the LCHF people. The LCHF movement ridicules the vegans. The vegans mock the Paleo supporters. The Paleo followers deride the low-fat devotees. All diets work because they all address a different aspect of the disease. But none of them work for very long, because none of them address the *totality* of the disease. Without understanding the *multifactorial* nature of obesity—which is critical—we are doomed to an endless cycle of blame.

Most dietary trials are fatally flawed by this tunnel vision. Trials comparing low-carb to low-calorie diets have all asked the wrong question. These two diets are not mutually exclusive. What if both are valid? Then there should be similar weight loss on both sides. Low-carb diets lower insulin. Lowering insulin levels reduces obesity. However, all foods raise insulin to some degree. Since refined carbohydrates often make up 50 per cent or more of the Standard American Diet, low-calorie diets generally result in a reduced carbohydrate intake. So low-calorie diets, by restricting the total amount of food consumed, still work to lower insulin levels. *Both will work – at least in the short term.*

That is exactly what Harvard professor Dr. Frank Sacks confirmed in his randomized study of four different diets.[3] Despite differences in carbohydrate, fat and protein content, albeit relatively minor, weight loss was the same. Maximum weight loss occurred at six months, with gradual regain thereafter. A 2014 meta-analysis of dietary trials reached much the same conclusion.[4] 'Weight loss differences between individual diets were minimal.' Sure, sometimes one diet comes off as slightly better than another. The difference is usually less than 2 pounds (about 1 kilogram) and often fades within a year. Let's face it. We've done low calories, low fat. It didn't work. We've done Atkins, too. It didn't produce the effortless weight loss that was promised.

Sometimes these results are interpreted to mean that everything can be eaten in moderation—which doesn't even begin to address the complexity of weight gain in humans. 'Moderation' is a cop-out answer—a deliberate attempt to evade the hard work of searching for dietary truths. For example, should we eat broccoli in the same moderation as ice cream? Obviously not. Should we drink milk in the same moderation as sugar-sweetened beverages? Obviously not. The long-recognized truth is that *certain* foods must be severely restricted, including sugar-sweetened beverages and candy. Other foods do not need to be restricted: kale or broccoli, for instance.

Others have erroneously concluded that 'it's all about calories.' Actually, it's nothing of the sort. Calories are only a single factor in the multifactorial disease that is obesity. Let's face the truth. Low-calorie diets have been tried again and again and again. They fail every single time.

There are other answers that are not really answers. These include, 'There is no best diet' or 'Choose the diet that suits you' or 'The best diet is one you can follow.' But if supposed experts in nutrition and disease don't know the right diet, how are you supposed to? Is the Standard American Diet is the best diet for me because it's the one I can follow? Or a diet of sugared cereals and pizza? Obviously not.

In cardiovascular disease, for example, 'Choose the treatment that suits you' would never be considered satisfactory advice. If the lifestyle factors of stopping smoking and increased physical activity both reduce heart disease, then we would strive to do both, rather than try to choose one or the other. We would *not* say, 'The best lifestyle for heart disease is the one you can follow.' Unfortunately, many so-called experts in obesity profess this exact sentiment.

The truth is that there are multiple overlapping pathways that lead to obesity. The common uniting theme is the hormonal imbalance of hyper-insulinemia. For some patients, sugar or refined carbohydrates are the main problem. Low-carbohydrate diets may work best here. For others, the main problem may be insulin resistance. Changing meal timing or intermittent fasting may be most beneficial. For still others, the cortisol pathway is dominant. Stress reduction techniques or correcting sleep deprivation may be critical. Lack of fiber may be the critical factor for yet others.

Most diets attack one part of the problem at a time. But why? In cancer treatment, for example, multiple types of chemotherapy and radiation are combined together. The probability of success is much higher with a broad-based attack. In cardiovascular disease, multiple drug treatments work together. We use drugs to treat high blood pressure, high cholesterol, diabetes and smoking cessation—all at the same time. Treating high blood pressure does not mean ignoring smoking. In challenging infections such as HIV, a cocktail of different antiviral medicines are combined together for maximum efficacy.

The same approach is necessary to address the multidimensional problem of obesity. Instead of targeting a single point in the obesity cascade, we need multiple targets and treatments. We don't need to choose sides. Rather than compare a dietary strategy of, say, low calorie versus low carb, why not do both? *There is no reason we can't.*

It is also important to tailor the approach individually to address the cause of the high insulin levels. For example, if chronic sleep deprivation is the main problem causing weight gain, then decreasing refined grains is not likely to help. If excessive sugar intake is the problem, then mindfulness meditation is not going to be especially useful.

Obesity is a hormonal disorder of fat regulation. Insulin is the major hormone that drives weight gain, so the rational therapy is to *lower insulin levels.* There are multiple ways to achieve this, and we should take advantage of each one. In the rest of this chapter, I will outline a step-by-step approach to accomplish this goal.

STEP 1: REDUCE YOUR CONSUMPTION OF ADDED SUGARS

SUGAR STIMULATES INSULIN secretion, but it is far more sinister than that. Sugar is particularly fattening because it increases insulin both immediately and over the long term. Sugar is comprised of equal amounts of glucose and fructose, as discussed in chapter 14, and fructose contributes directly to insulin resistance in the liver. Over time, insulin resistance leads to higher insulin levels.

Therefore, sucrose and high fructose corn syrup are exceptionally fattening, far in excess of other foods. Sugar is uniquely fattening because it directly produces insulin resistance. With no redeeming nutritional qualities, added sugars are usually one of the first foods to be eliminated in *any* diet.

Many natural, unprocessed whole foods contain sugar. For example, fruit contains fructose, and milk contains lactose. Naturally occurring and added sugars are distinct from one another. The two key differences between them are amount and concentration.

Obviously, first you should remove your sugar bowl from your table. There is no reason to add sugar to any food or beverage. But sugars are often hidden in the preparation of food, which means that avoiding sugar is often difficult and you can ingest a surprisingly large amount without knowing it. Sugars are often added to foods during processing or cooking, which presents dieters with several potential pitfalls. First, sugars may be added in unlimited amounts. Second, sugar may be present in processed food in much higher concentrations than in natural foods. Some processed foods are virtually 100 per cent sugar. This condition almost does not exist in natural foods, with honey possibly being the exception. Candy is often little more than flavored sugar. Third, sugar may be ingested by itself, which may cause people to overeat sugary treats, as there is nothing else within the food to make you 'full.' There is often no dietary fiber to help offset the harmful effects. For these reasons, we direct most of our efforts toward reducing added, rather than natural sugars in our diet.

Read the labels

Almost ubiquitous in refined and processed foods, sugar is not always labeled as such. Other names include sucrose, glucose, fructose, maltose, dextrose, molasses, hydrolyzed starch, honey, invert sugar, cane sugar, glucose-fructose, high fructose corn syrup, brown sugar, corn sweetener, rice/corn/cane/maple/malt/golden/palm syrup and agave nectar. These aliases attempt to conceal the presence of large amounts of added sugars. A popular trick is to use several different pseudonyms on the food's label. This trick prevents 'sugar' from being listed as the first ingredient.

The addition of sugar to processed foods provides almost magical flavor-enhancing properties at virtually no cost. Sauces are serial offenders. Barbeque, plum, honey garlic, hoisin, sweet and sour, and other dipping sauces contain large amounts of sugar. Spaghetti sauce may contain as much as 10 to 15 grams of sugar (3 to 4 teaspoons). This counters the tartness of the tomatoes, and therefore may not be immediately evident to your taste buds. Commercial salad dressings and condiments such as ketchup and relish often contain lots of sugar. The bottom line is this: If it comes in a package, it probably contains added sugar.

Asking how much sugar is acceptable is like asking how many cigarettes are acceptable. Ideally, no added sugar at all would be best, but that probably will not happen. Still, see the next section for some reasonable suggestions.

What to do about dessert

Most desserts are easily identified and eliminated from your diet. Desserts are mostly sugar with complementary flavors added. Examples include cakes, puddings, cookies, pies, mousses, ice cream, sorbets, candy and candy bars.

So what can you do about dessert? Follow the example of traditional societies. The best desserts are fresh seasonal fruits, preferably locally grown. A bowl of seasonal berries or cherries with whipped cream is a delicious way to end a meal. Alternatively, a small plate of nuts and cheeses also makes for a very satisfying end to a meal, without the burden of added sugars.

Dark chocolate with more than 70 per cent cacao, in moderation, is a surprisingly healthy treat. The chocolate itself is made from cocoa beans and does not naturally contain sugar. (However, most milk chocolate *does* contain large amounts of sugar.) Dark and semisweet chocolate contain less sugar than milk or white varieties. Dark chocolate also contains significant amounts of fiber and antioxidants such as polyphenols and flavanols. Studies on dark-chocolate consumption indicate that it may help reduce blood pressure,[5] insulin resistance[6] and heart disease.[7] Most milk chocolates, by contrast, are little more than candies. The cacao component is too small to be beneficial.

Nuts, in moderation, are another good choice for an after-dinner indulgence. Most nuts are full of healthful monounsaturated fats, have little or no carbohydrates, and are also high in fiber, which increases their potential benefit. Macadamia nuts, cashews and walnuts can all be enjoyed. Many studies show an association between increased nut consumption and better health, including reducing heart disease[8] and diabetes.[9] Pistachio nuts, high in the antioxidant gamma-tocopherol and vitamins such as manganese, calcium, magnesium and selenium, are widely eaten in the Mediterranean diet. A recent Spanish study found that adding 100 pistachios to one's daily diet improved fasting glucose, insulin and insulin resistance.[10]

That is not to say that sugar cannot be an occasional indulgence. Food has always played a major role in celebrations—birthdays, weddings, graduations, Christmas, Thanksgiving, etc. The key word here is *occasional.* Dessert is not to be taken every day.

Be aware, though, that if your goal is weight loss, your first major step must be to severely restrict sugar. Don't replace sugar with artificial sweeteners, as they also raise insulin as much as sugar and are equally prone to causing obesity. (See chapter 15.)

Just don't snack

The healthy snack is one of the greatest weight-loss deceptions. The myth that 'grazing is healthy' has attained legendary status. If we were meant to 'graze,' we would be cows. Grazing is the direct opposite of virtually all food traditions. Even as recently as the 1960s, most people still ate just three meals per day. Constant stimulation of insulin eventually leads to insulin resistance. (For more on the dangers of snacking, see chapters 10 and 11.)

The solution? Stop eating all the time.

Snacks are often little more than thinly disguised desserts. Most contain prodigious amounts of refined flour and sugar. These pre-packaged conveniences have taken over the supermarket shelves. Cookies, muffins, pudding, Jell-O, fruit roll-ups, fruit leather, chocolate bars, cereal bars, granola bars and biscuits—all are best avoided. Rice cakes, advertising themselves as low fat, compensate for lack of taste with sugar. Canned or processed fruit conceals buckets of sugar behind the healthy image of the fruit. A serving of Mott's Applesauce contains 5½ teaspoons of sugar (22 grams). A serving of canned peaches contains 4½ teaspoons of sugar (18 grams).

Are snacks necessary? No. Simply ask yourself this question. Are you really hungry or just bored? Keep snacks completely out of sight. If you have a snack habit, replace that habit loop with one that is less destructive to your health. Perhaps a cup of green tea in the afternoon should be your new habit. There's a simple answer to the question of what to eat at snack time. Nothing. Don't eat snacks. Period. Simplify your life.

Make breakfast optional

Breakfast is, without question, the most controversial meal of the day. The advice to eat something, anything, as soon as you step out of bed is often heard. But breakfast needs to be downgraded from 'most important meal of the day' to 'meal.' Different nations have different breakfast traditions. The big 'American' breakfast contrasts directly with the French 'petit dejeuner' or 'small lunch.' The key word here is 'small.'

The greatest problem is that, like snacks, breakfast foods are often little more than dessert in disguise, containing vast quantities of highly processed carbohydrates and sugar. Breakfast cereals, particularly those that target children, are among the worst offenders. On average, they contain 40 per cent more sugar than those that target adults.[11] Not surprisingly, almost all cereals for children contain sugar, and ten contain more than 50 per cent sugar by weight. Only 5.5 per cent met the standard for 'low sugar.' In the diets of children under age eight, breakfast cereals rank behind only candy, cookies, ice cream and sugared drinks as a source of dietary sugar.

A simple rule to follow is this: Don't eat sugared breakfast cereal. If you must, eat cereals containing less than 0.8 of a teaspoon (4 grams) of sugar per serving.

Many breakfast items from the bakery are also highly problematic: muffins, cakes, Danishes and banana bread. Not only do they contain significant amounts of refined carbohydrates, they are often sweetened with sugars and jams. Bread often contains sugar and is eaten with sugary jams and jellies. Peanut butter often contains added sugars, too.

Traditional and Greek yogurts are nutritious foods. However, commercial yogurts are made with large amounts of added sugars and fruit flavorings. A serving of Yoplait fruit yogurt contains almost 8 teaspoons of sugar (31 grams). Oatmeal is another traditional and healthy food. Whole oats and steel-cut oats are a good choice, requiring long cooking times because they contain significant amounts of fiber that requires heat and time to break down. Avoid instant oatmeal. It is heavily processed and refined, which allows for instant cooking, and it contains large amounts of added sugar and flavors. Most of the nutritional content is gone. Quaker's flavored instant oatmeal may contain up to 3¼ teaspoons of sugar (13 grams) of sugar per serving. Instant cream of wheat has the same problem. A single serving contains 4 teaspoons (16 grams) of sugar. While rolled oats and dried fruit, granola and granola bars attempt to disguise themselves as healthy, they are often heavily sugared and contain chocolate chips or marshmallows.

Eggs, previously shunned due to cholesterol concerns, can be enjoyed in a variety of ways: scrambled, over easy, sunny side up, hard-boiled, soft-boiled, poached, etc. Egg whites are high in protein, and the yolk contains many vitamins and minerals, including choline and selenium. Eggs are particularly good sources of lutein and zeaxanthin, antioxidants that may help protect against eye problems such as macular degeneration and cataracts.[12] The cholesterol in eggs may actually help your cholesterol profile by changing cholesterol particles to the larger, less atherogenic particles.[13] Indeed, large epidemiologic studies have failed to link increased egg consumption to increased heart disease.[14,15] Most of all, eat eggs because they are delicious, whole, unprocessed foods.

In thinking about what to eat for breakfast, consider this: If you are not hungry, don't eat anything at all. It's perfectly acceptable to break your fast at noon with grilled salmon and a side salad. But there's nothing inherently wrong with eating breakfast in the morning either. It is just like any other meal. However, in the morning rush, many people tend to reach for conveniently prepackaged, heavily processed and heavily sugared foods. Eat whole, unprocessed foods at all meals, including breakfast. And if you don't have time to eat? Then don't eat. Again, simplify your life.

Beverages: No sugar added

The sugar-sweetened drink is one of the leading sources of added sugars. This includes all soda pop, sugar-sweetened teas, fruit juice, fruit punch, vitamin water, smoothies, shakes, lemonade, chocolate or flavored milk, iced coffee drinks and energy drinks. Hot drinks such as hot chocolate, mochaccino, caffè mocha and sweetened coffee and tea can also be included. Trendy alcoholic drinks add significant amounts of sugar to your diet, including drinks such as 'hard' lemonade, flavored wine coolers, cider beers as well as more traditional drinks such as Baileys Irish Cream, margaritas, daiquiris, piña coladas, dessert wines, ice wines, sweet sherries and liqueurs.

What about alcohol itself? Alcohol is made from the fermentation of sugars and starches from various sources. Yeast eat the sugars and convert them to alcohol. Residual sugars result in a sweeter beverage. Sweetened dessert wines are obviously full of sugar and are not recommended.

However, moderate consumption of red wine does not raise insulin or impair insulin sensitivity, and therefore may be enjoyed.[16] Up to two glasses a day is not associated with major weight gain[17] and may improve insulin sensitivity.[18] The alcohol itself, even from beer, seems to have minimal effects on insulin secretion or insulin resistance. It is sometimes said that you get fat from the foods you eat with the alcohol rather than from the alcohol itself. There may be some truth to that, although the evidence is sparse.

So what is left to drink? The best drink is really just plain or sparkling water. Slices of lemon, orange or cucumber are a refreshing addition. Several traditional and delicious drinks are also available as described below.

Coffee: Healthier than we thought

Due to its high caffeine content, coffee is sometimes considered unhealthy. However, recent research has come to the opposite conclusion,[19] perhaps due to the fact that coffee is a major source of antioxidants,[20] magnesium, lignans[21] and chlorogenic acid.[22]

Coffee, even the decaffeinated version, appears to protect against type 2 diabetes. In a 2009 review, each additional daily cup of coffee lowered the risk of diabetes by 7 per cent, even up to six cups per day.[23] The European Prospective Investigation into Cancer and Nutrition study estimated that drinking at least three cups of coffee or tea daily reduced the risk of diabetes by 42 per cent.[24] The Singapore Chinese Health Study showed a 30 per cent reduction in risk.[25]

Coffee drinking is associated with a 10 per cent to 15 per cent reduction in total mortality.[26] Large-scale studies[27] found that most major causes of death, including heart disease, were reduced. Coffee may guard against the neurologic diseases Alzheimer's,[28,29] Parkinson's disease,[30,31] liver cirrhosis[32] and liver cancer.[33] A word of caution here: While these correlation studies are suggestive, they are not proof of benefit. However, they suggest that coffee may not be as harmful as we imagined.

Store beans in an airtight container away from excessive moisture, heat and light. Flavor is lost quickly after grinding, so investing in a reliable grinder is worthwhile. Grind beans immediately before brewing. On hot days, iced coffee is simple and inexpensive to make. Simply brew a pot of regular coffee and cool it in the refrigerator overnight. You can use cinnamon, coconut oil, vanilla extract, almond extract and cream to flavor your coffee without changing its healthy nature. Avoid adding sugar or other sweeteners.

Teatime, anytime

After water, tea is the most popular beverage in the world. There are several basic tea varieties. Black tea is the most common, making up almost 75 per cent of global consumption. Harvested leaves are fully fermented, giving tea its characteristic black color. Black tea tends to be higher in caffeine than other varieties. Oolong tea is semi-fermented, meaning that it undergoes a shorter period of fermentation. Green tea is non-fermented. Instead, the freshly harvested leaves immediately undergo a steaming process to stop fermentation, giving green tea a much more delicate and floral taste. Green tea is naturally much lower in caffeine than coffee, making this drink ideal for those who are sensitive to caffeine's stimulant effects.

Green tea contains large concentrations of a group of powerful antioxidants called catechins; notably one called epigallocatechin-3-gallate. Catechins may play a role in inhibiting carbohydrate-digestive enzymes, resulting in lower glucose levels[34] and protecting the pancreatic beta cells.[35] Fermentation (black tea) changes the catechins to a variety of theaflavins,[36] making the antioxidant potential of green tea and black tea comparable. Polyphenols in green tea are also believed to boost metabolism,[37] which may aid in fat burning.[38] Many health benefits have been ascribed to green-tea consumption, including increased fat oxidation during exercise,[39] increased resting energy expenditure,[40] lower risk of various types of cancer.[41]

A meta-analysis of studies confirms that green tea helps with weight loss, although the benefit is rather modest: in the range of about 2 to 4 pounds (1 to 2 kilograms).[42] Studies, including the Singapore Chinese Health Study, showed that tea drinking reduced the risk of type 2 diabetes by 14 to 18 per cent.[43, 44]

All teas may be enjoyed both as hot or cold beverages. There are infinite varieties of tea available to suit any taste. Flavor can be added with the addition of lemon peel, orange peel, cinnamon, cardamom, vanilla pods, mint and ginger.

Herbal teas are infusions of herbs, spices or other plant matter in hot water. These are not true teas since they do not contain tea leaves. They make excellent drinks without added sugars, and can be enjoyed hot or cold. The varieties are endless. Some popular varieties include mint, chamomile, ginger, lavender, lemon balm, hibiscus and rosehip teas. The addition of cinnamon or other spices can enhance the flavor.

Bone broth

Virtually every culture's culinary traditions include the nutritious and delicious bone broth. Animal bones are simmered with the addition of vegetables, herbs and spices for flavoring. The long simmering time (four to forty-eight hours) releases most of the minerals, gelatin and nutrients. The addition of a small amount of vinegar during cooking helps leach some of the stored minerals. Bone broths are very high in amino acids such as proline, arginine and glycine, as well as minerals such as calcium, magnesium and phosphorus.

Animal bones are often available at ethnic grocery stores and fairly inexpensive. They are also very convenient, requiring little preparation time. They can be made in large batches and frozen. Most commercially prepared broths have nothing in common with the homemade variety. Prepackaged broths often rely on artificial flavors and MSG to provide taste. The minerals, nutrients and gelatin are not present in many canned broths.

STEP 2: REDUCE YOUR CONSUMPTION OF REFINED GRAINS

REFINED GRAINS SUCH as white flour stimulate insulin to a greater degree than virtually any other food. If you reduce your consumption of flour and refined grains, you will substantially improve your weight-loss potential. White flour, being nutritionally bankrupt, can be safely reduced or even eliminated from your diet. Enriched white flours have had all their nutrients stripped out during processing and later added back to retain a veneer of healthiness.

Whole wheat and whole grains are an improvement over white flour, containing more vitamins and fiber. The bran fiber helps protect against the insulin spikes. However, whole-grain flour is still highly processed in a modern flourmill. Traditional stone-mill grinding is preferable. The ultrafine particles produced by modern milling techniques ensure rapid absorption of flour, even whole-wheat flour, by the intestine, which tends to increase the insulin effect.

Avoid processed bakery foods that are mostly flour and other starches: bread, bagels, English muffins, roti, naan breads, dinner rolls, bread sticks, Melba toasts, crackers, tea biscuits, scones, tortillas, wraps, muffins, cookies, cakes, cupcakes and donuts. Pasta and noodles of all varieties are also concentrated sources of refined carbohydrates; reduce these to a minimum. The whole-grain pastas that are now widely available are a better choice, though far from ideal.

Carbohydrates should be enjoyed in their natural, whole, unprocessed form. Many traditional diets built around carbohydrates cause neither poor health nor obesity. Remember: the toxicity in much Western food lies in the processing, rather than in the food itself. The carbohydrates in Western diets are heavily skewed toward refined grains, and are thus highly obesogenic. Eggplant, kale, spinach, carrots, broccoli, peas, Brussels sprouts, tomatoes, asparagus, bell peppers, zucchini, cauliflower, avocados, lettuce, beets, cucumbers, watercress, cabbage, among others, are all extremely healthy carbohydrate-containing foods.

Quinoa, technically a seed but often used as a grain, is one of the so-called ancient grains. Grown originally in the Inca empire of South America, it was referred to as the 'mother of all grains.' It comes in three varieties: red, white and black. Quinoa is very high in fiber, protein and vitamins. In addition, quinoa has a low glycemic index and contains plenty of antioxidants, such as quercetin and kaempferol, which are believed to be anti-inflammatory.

Chia seeds are native to South and Central America and have been dated to the Aztec and Mayans. The word chia is derived from the ancient Mayan word for strength. Chia seeds are high in fiber, vitamins, minerals, omega 3, proteins and antioxidants. They are usually soaked in fluid, as they absorb ten times their weight in water, forming an edible gel.

Beans are a versatile, fiber-rich carbohydrate staple of many traditional diets. They are an extremely good source of protein, particularly for vegetarian diets. Edamame beans, popular in Japanese cuisine, provide 9 grams of fiber and 11 grams of protein per serving.

STEP 3: MODERATE YOUR PROTEIN CONSUMPTION

IN CONTRAST TO refined grains, protein cannot and should not be eliminated from your diet. (For more on protein, see chapter 17.) Instead, moderate the amount of protein in your diet to fall within 20 per cent to 30 per cent of your total calories.

Excessively high-protein diets are not advisable and are quite difficult to follow, since protein is rarely eaten in isolation. Protein-containing foods such as dairy or meat often contain significant amounts of fat. Vegetable proteins, such as legumes, often contain significant amounts of carbohydrate. Thus, extremely high-protein diets are usually quite unpalatable. They tend to rely on egg whites and very lean meats. Needless to say, it's difficult to comply with such very limited diets. Some dieters turn to meal replacement shakes, bars or protein powders, which are really just highly processed 'fake foods.' Optifast, Slim-Fast, Ensure and Boost are only some examples in a crowded marketplace of nutritional thieves. These products don't produce lasting weight loss and they're designed to keep you hooked on their processed concoctions.

STEP 4: INCREASE YOUR CONSUMPTION OF NATURAL FATS

OF THE THREE major macronutrients (carbohydrates, proteins and fats), dietary fat is the least likely to stimulate insulin. Thus, dietary fat is not inherently fattening, but potentially protective. (For more about fat as a protective factor, see chapter 18.) In choosing fats, strive for a higher proportion of natural fats. Natural, unprocessed fats include olive oil, butter, coconut oil, beef tallow and leaf lard. The highly processed vegetable oils, high in inflammatory omega 6 fatty acids, may have some detrimental health effects.

Widely acknowledged as healthy, the Mediterranean diet is high in oleic acid, the monounsaturated fats contained in olive oil. Olives are native to the Mediterranean region, and olive oil was being produced as early as 4500 BC. Ripe olive fruit is crushed into a paste and the oil is extracted using a press. The term 'virgin' refers to oil that is extracted using these mechanical means only and is certainly the best choice. Other grades of oil rely on chemical methods and should be avoided. 'Refined' oils use chemicals and high heat to extract the oil and neutralize bad tastes, allowing producers to use second-rate olives. Be aware that the term 'pure olive oil' often denotes refined oils. Extra-virgin olive oil is unrefined, contains fruity undertones, and it meets certain quality standards.

The health benefits of olive oil have long been recognized. Olive oil contains large amounts of antioxidants including polyphenols and oleocanthal,[45] which has anti-inflammatory properties. Among its purported benefits are reduced inflammation, lowered cholesterol,[46] decreased blood clotting[47] and reduced blood pressure.[48] Together, these potential properties may reduce the overall risk of cardiovascular disease, including heart attacks and strokes.[49]

Heat and light cause oxidation, so olive oil must be stored in a cool, dark spot. Dark-green glass containers reduce incoming light to help preserve the oil. Light olive oils undergo a fine filtration to remove most of the flavor, aroma and color. This process makes it more suitable for baking, where the fruity aroma is not desirable.

Nuts are also prominent in the Mediterranean diet. Long shunned for their high fat content, they have since been recognized to have significant health benefits. In addition to healthy fats, nuts are naturally high in fiber and low in carbohydrates. Walnuts in particular are high in the omega 3 fatty acids.

Full-fat dairy is delicious and can be enjoyed without concern of fattening effects. A review of twenty-nine randomized control trials[50] showed neither a fat-gaining nor reducing effect. Full-fat dairy is associated with a 62 per cent lower risk of type 2 diabetes.[51]

Avocados have been recently recognized as a very healthy and delicious addition to any diet. Although not sweet, they are the fruit of the avocado tree. High in vitamins and particularly high in potassium, the avocado is unique among fruits for being very low in carbohydrates and high in the monounsaturated fat oleic acid. Furthermore, it is very high in both soluble and insoluble fiber.

STEP 5: INCREASE YOUR CONSUMPTION OF PROTECTIVE FACTORS

FIBER CAN REDUCE the insulin-stimulating effects of carbohydrates, making it one of the main protective factors against obesity, but the average North American diet falls far short of recommended daily intakes. (For more on fiber as a protective factor, see chapter 16.) Numerous studies and observations have confirmed the weight-lowering effects of dietary fiber. Natural whole foods contain plenty of fiber, which is often removed during processing. Fruits, berries, vegetables, whole grains, flax seeds, chia seeds, beans, popcorn, nuts, oatmeal and pumpkin seeds provide ample fiber.

Glucomannan is a soluble, fermentable and highly viscous dietary fiber that comes from the root of the elephant yam, also known as konjac, native to Asia. Glucomannan can absorb up to fifty times its weight in water, making it one of the most viscous dietary fibers known.[52] The konjac tuber has been used for centuries as a herbal remedy and to make traditional foods such as konjac jelly, tofu and noodles.

Vinegar is also a protective factor. Used in many traditional foods, it may help reduce insulin spikes. Italians often eat bread dipped in oil and vinegar—a prime example of eating a high-carb food with protective factors. Vinegar is added to sushi rice, which reduces its glycemic index by 20 per cent to 40 per cent.[53] Fish and chips are often eaten with malt vinegar. Apple cider vinegar may be taken diluted in some water.

THE LAST PIECE OF THE PUZZLE

THERE ARE FIVE basic steps in weight loss:

1. Reduce your consumption of added sugars.
2. Reduced your consumption of refined grains.
3. Moderate your protein intake.
4. Increase your consumption of natural fats.
5. Increase your consumption of fiber and vinegar.

When it comes to the question of what to eat, you pretty much already knew the answer. Most diets very conspicuously resemble each other. There is far more agreement than discord. Eliminate sugars and refined grains. Eat more fiber. Eat vegetables. Eat organic. Eat more home-cooked meals. Avoid fast food. Eat whole unprocessed foods. Avoid artificial colors and flavors. Avoid processed or microwavable foods. Whether you follow the low-carb, low-calorie, South Beach, Atkins or some other mainstream diet, the advice is very similar. Sure, there are particular nuances to each diet, particularly with respect to dietary fats, but they tend to agree more than they disagree. So, why all the controversy?

Agreement does not sell books or magazine. We always need to 'discover' the latest and greatest 'superfood.' Acai berries. Quinoa. Or we need to 'discover' the latest and greatest dietary villain. Sugar. Wheat. Fat. Carbohydrates. Calories. *Vogue* magazine does not carry headlines such as 'Diet advice you already knew!'

All diets work in the short term. But we've been ignoring the long-term problem of insulin resistance. There is one more piece of the puzzle—a solution found many centuries ago. A practice that has been enshrined in the nutritional lore of virtually every population on earth. A tradition rapidly becoming extinct.

This tradition is the subject of the next chapter.

(20)

WHEN TO EAT

•

There is nothing new, except what has been forgotten.
MARIE ANTOINETTE

LONG-TERM DIETING IS futile. After the initial weight loss, the dreaded plateau appears, followed by the even more dreaded weight regain. The body reacts to weight loss by trying to return to its original body set weight. We hope our body set weight will decrease over time, but that hoped-for reduction does not materialize. Even if we eat all the right things, our insulin levels stay elevated.

But we've been addressing only half of the problem. Long-term weight loss is really a two-step process. *Two* major factors maintain our insulin at a high level. The first is the foods that we eat—which are what we usually change when we go on a diet. But we fail to address the other factor: the long-term problem of insulin resistance. This problem is one of meal timing.

Insulin resistance keeps our insulin levels high. High insulin maintains our high body set weight. Inexorably, our high body set weight erodes our weight-loss efforts. We start feeling hungrier. Our metabolism (that is, our total energy expenditure) relentlessly decreases until it falls below the level of our energy intake. Our weight plateaus and ruthlessly climbs back up to our original body set weight, even as we keep dieting. Clearly, changing what we eat is not always enough.

To succeed, we must break the insulin-resistance cycle. But how? The body's knee-jerk reaction to insulin resistance is to increase insulin levels, which, in turn, creates even more resistance. To break the insulin-resistance cycle, we must have recurrent periods of very *low* insulin levels. (Remember that resistance depends on having both persistent and high levels.)

But how can we induce our body into a temporary state of very low insulin levels?

We know that eating the proper foods prevents high levels, but it won't do much to lower them. Some foods are better than others; nonetheless, all foods increase insulin production. If all foods raise insulin, then the only way for us to lower it is to completely abstain from food.

The answer we are looking for is, in a word, fasting.

When we talk about fasting to break insulin resistance and lose weight, we are talking about intermittent fasts of twenty-four to thirty-six hours. A practical plan for accomplishing these fasts is included in appendix B. The remainder of this chapter will be devoted to addressing the health concerns around fasting—which, the research shows us, is a beneficial practice.

FASTING: AN ANCIENT REMEDY

INSTEAD OF SEARCHING for some exotic, never-seen-before diet miracle to help us break insulin resistance, let's instead focus on a tried-and-true ancient healing tradition. Fasting is one of the oldest remedies in human history and has been part of the practice of virtually every culture and religion on earth.

Whenever fasting is mentioned, there is always the same eye-rolling response: Starvation? That's the answer? No. Fasting is completely different. Starvation is the *involuntary* absence of food. It is neither deliberate nor controlled. Starving people have no idea when and where their next meal will come from. Fasting, however, is the *voluntary* abstinence from food for spiritual, health or other reasons. Fasting may be done for any period of time, from a few hours to a few months. In a sense, fasting is part of everyday life. The term 'breakfast' is the meal that breaks the fast—which we do daily.

As a healing tradition, fasting has a long history. Hippocrates of Kos (c. 460–c. 370 BC) is widely considered the father of modern medicine. Among the treatments that he prescribed and championed were the practice of fasting and the consumption of apple cider vinegar. Hippocrates wrote, 'To eat when you are sick, is to feed your illness.' The ancient Greek writer and historian Plutarch (c. AD 46–c. AD 120) also echoed these sentiments. He wrote, 'Instead of using medicine, better fast today.' Plato and his student Aristotle were also staunch supporters of fasting.

The ancient Greeks believed that medical treatment could be discovered by observing nature. Humans, like most animals, do not eat when they become sick. Consider the last time you were sick with the flu. Probably the last thing you wanted to do was eat. Fasting seems to be a universal human response to multiple forms of illnesses and is ingrained in human heritage, as old as mankind itself. Fasting is, in a sense, an instinct.

The ancient Greeks believed that fasting improved cognitive abilities. Think about the last time you ate a huge Thanksgiving meal. Did you feel more energetic and mentally alert afterward? Or instead, did you feel sleepy and a little dopey? More likely the latter. Blood is shunted to your digestive system to cope with the huge influx of food, leaving less blood for brain function. Fasting does the opposite, leaving more blood for your brain.

Other intellectual giants were also great proponents of fasting. Paracelsus (1493–1541), the founder of toxicology and one of the three fathers of modern Western medicine (along with Hippocrates and Galen), wrote, 'Fasting is the greatest remedy—the physician within.'

Benjamin Franklin (1706–90), one of America's founding fathers and renowned for wide knowledge, once wrote of fasting, 'The best of all medicines is resting and fasting.'

Fasting for spiritual purposes is widely practiced and remains part of virtually every major religion in the world. Jesus Christ, Buddha and the prophet Muhammed all shared a common belief in the power of fasting. In spiritual terms, it is often called cleansing or purification; in practical terms, it amounts to the same thing. The practice of fasting developed independently among different religions and cultures, not as something that was harmful, but something that was deeply, intrinsically beneficial to the human body and spirit.[1] In Buddhism, food is often consumed only in the morning, and followers fast daily from noon until the next morning. In addition to this, there may be various water-only fasts for days or weeks on end. Greek Orthodox Christians may follow various fasts over 180–200 days of the year. Dr. Ancel Keys often considered Crete the poster child for the healthy Mediterranean diet. However, there was a critically important factor that he completely dismissed. Most of the population of Crete followed the Greek Orthodox tradition of fasting.

Muslims fast from sunrise to sunset during the holy month of Ramadan. The prophet Muhammad also encouraged fasting every week on Mondays and Thursdays. Ramadan differs from many fasting protocols in that fluids, in addition to food, are forbidden, so practitioners of this particular fast undergo a period of mild dehydration. Further, since eating is permitted before sunrise and after sunset, recent studies[2] indicate that daily caloric intake actually rises significantly during this period. Gorging, particularly on highly refined carbohydrates, before sunrise and after sunset negates much of fasting's benefit.

THE BODY'S RESPONSE TO FASTING

GLUCOSE AND FAT are the body's main sources of energy. When glucose is not available, then the body adjusts by using fat, without any health detriment. This compensation is a natural part of life. Periodic food scarcity has always been part of human history, and our bodies have evolved processes to deal with this fact of Paleolithic life. The transition from the fed state to the fasted state occurs in several stages:[3]

1. *Feeding:* During meals, insulin levels are raised. This allows glucose uptake by tissues such as the muscle or brain for direct use as energy. Excess glucose is stored as glycogen in the liver.
2. *The post-absorptive phase* (six to twenty-four hours after fasting starts): Insulin levels being to fall. The breakdown of glycogen releases glucose for energy. Glycogen stores last for roughly twenty-four hours.
3. *Gluconeogenesis* (twenty-four hours to two days): The liver manufactures new glucose from amino acids and glycerol. In non-diabetic persons, glucose levels fall but stay within the normal range.
4. *Ketosis* (one to three days after fasting starts): The storage form of fat, triglycerides, is broken into the glycerol backbone and three fatty acid chains. Glycerol is used for gluconeogenesis. Fatty acids may be used for directly for energy by many tissues in the body, but not the brain. Ketone bodies, capable of crossing the blood-brain barrier, are produced from fatty acids for use by the brain. Ketones can supply up to 75 per cent of the energy used by the brain.[4] The two major types of ketones produced are beta hydroxybutyrate and acetoacetate, which can increase more than seventy-fold during fasting.[5]
5. *Protein conservation phase* (after five days): High levels of growth hormone maintain muscle mass and lean tissues. The energy for maintenance of basal metabolism is almost entirely met by the use of free fatty acids and ketones. Increased norepinephrine (adrenalin) levels prevent the decrease in metabolic rate.

The human body is well adapted for dealing with the absence of food. What we're describing here is the process the body undergoes to switch from burning glucose (short term) to burning fat (long term). Fat is simply the body's stored food energy. In times of food scarcity, stored food (fat) is naturally released to fill the void. The body does not 'burn muscle' in an effort to feed itself until all the fat stores are used.

It's crucial to note that all these beneficial adaptive changes do not occur in the caloric-reduction diet strategy.

HOW YOUR HORMONES ADAPT TO FASTING

Insulin

Fasting is the most efficient and consistent strategy to decrease insulin levels, a fact first noted decades ago[6] and widely accepted as true. All foods raise insulin; therefore, the most effective method of reducing insulin is to avoid all foods. Blood glucose levels remain normal as the body switches over to burning fat for energy. This effect occurs with fasting periods as short as twenty-four to thirty-six hours. Longer fasts reduce insulin even more dramatically. More recently, alternate daily fasting has been studied as an acceptable technique for reducing insulin levels.[7]

Regular fasting, by routinely lowering insulin levels, has been shown to significantly improve insulin sensitivity.[8] *This finding is the missing piece in the weight-loss puzzle.* Most diets restrict the intake of foods that cause increased insulin secretion, but don't address insulin resistance. You lose weight initially, but insulin resistance keeps your insulin levels and body set weight high. By fasting, you can efficiently reduce your body's insulin resistance, since it requires both persistent and high levels.

Insulin causes salt and water retention in the kidney, so lowering insulin levels rids the body of excess salt and water. Fasting is often accompanied by an early, rapid weight loss. For the first five days, weight loss averages 1.9 pounds (0.9 kilograms) per day, far exceeding the loss that could be expected from the caloric restriction, and is probably due to diuresis. Diuresis reduces bloating and may also lower blood pressure slightly.

Growth hormone

Growth hormone is known to increase the availability and utility of fats for fuel. It also helps to preserve muscle mass and bone density.[9] Growth hormone secretion is difficult to measure accurately because of its intermittent release, but it decreases steadily with age. One of the most potent stimuli to growth hormone secretion is fasting.[10] Over a five-day fasting period, growth hormone secretion more than doubled. The net physiologic effect is to maintain muscle and bone tissue mass during the fasting period.

Adrenalin

Fasting increases adrenalin levels, starting at around twenty-four hours. Forty-eight hours of fasting produces a 3.6 per cent increase in metabolic rate,[15] not the dreaded metabolic shutdown so often seen in caloric-reduction strategies. In response to a four-day fast,[16] resting energy expenditure increased up to 14 per cent. Rather than slowing metabolism, the body revs it up instead. Presumably, it does this so we have energy to go out and find more food.

Electrolytes

Many people worry that fasting may cause malnutrition, but this concern is misplaced. The body's fat stores are, for most of us, quite ample for our bodies' needs. Even studies of prolonged fasting have found no evidence of malnutrition or micronutrient deficiency. Potassium levels may decrease slightly, but even two months of continuous fasting did not decrease levels below normal, even without the use of supplements.[11] Note that this duration of fasting is far longer than is generally recommended without medical supervision.

Magnesium, calcium and phosphorus levels during fasting are stable[12]—presumably, because of the large stores of these minerals in the bones. Ninety-nine per cent of the body's calcium and phosphorus is stored in the bones. A multivitamin supplement will provide the recommended daily allowance of micronutrients. In one case, a therapeutic fast of 382 days was maintained with only a multivitamin, with no harmful effect on the subject's health. Actually, this man maintained that he felt terrific during the entire period.[13] There were no episodes of hypoglycemia, as blood sugars were maintained within the normal range. The only concern may be a slight elevation in uric acid, which has been described in fasting.[14]

MYTHS ABOUT FASTING

MANY FASTING MYTHS have been repeated so often that they are often perceived as infallible truths. Consider the following:

- Fasting will make you lose muscle/burn protein.
- The brain needs glucose to function.
- Fasting puts you in starvation mode/lowers basal metabolism.
- Fasting will overwhelm you with hunger.
- Fasting causes overeating when you resume feeding.
- Fasting deprives the body of nutrients.
- Fasting causes hypoglycemia.
- It's just crazy.

If these myths were true, none of us would be alive today. Think about the consequences of burning muscle for energy. During long winters, there were many days where no food was available. After the first episode, you would be severely weakened. After several repeated episodes, you would be so weak that you would be unable to hunt or gather food. Humans would never have survived as a species. The better question would be why the human body would store energy as fat if it planned to burn protein instead. The answer, of course, is that is does not burn muscle in the absence of food. That is only a myth.

Starvation mode, as it is popularly known, is the mysterious bogeyman always raised to scare us away from missing even a single meal. This is simply absurd. Breakdown of muscle tissue happens only at extremely low levels of body fat—approximately 4 per cent—which is not something most people need to worry about. At this point, there is no further body fat to be mobilized for energy, and lean tissue is consumed. The human body has evolved to survive episodic periods of starvation. Fat is stored energy and muscle is functional tissue. Fat is burned first. This situation is akin to storing a huge amount of firewood but deciding to burn your sofa instead. It's stupid. Why do we assume the human body is so stupid? The body preserves muscle mass until fat stores become so low that it has no other choice.

Studies of alternate daily fasting, for example, show that the concern over muscle loss is largely misplaced.[17] Alternate daily fasting over seventy days decreased body weight by 6 per cent, but fat mass decreased by 11.4 per cent. Lean mass (including muscle and bone) did not change at all. Significant improvements were seen in LDL cholesterol and triglyceride levels. Growth hormone increased to maintain muscle mass. Studies of eating a single meal per day[18] found significantly more fat loss, compared to eating three meals per day, despite the same caloric intake. Significantly, no evidence of muscle loss was found.

There is another persistent myth that brain cells require glucose for proper functioning. This is incorrect. Human brains, unique among animals, can use ketones as a major fuel source during prolonged starvation, allowing the conservation of protein such as skeletal muscle. Again, consider the consequences if glucose were absolutely necessary for survival: humans just wouldn't survive. After twenty-four hours, glucose becomes depleted. If our brains had no alternative, we would become blubbering idiots as our brains shut down. Our intellect, our only advantage against wild animals, would begin to disappear. Fat is the body's way of storing food energy for the long term; it uses glucose/glycogen in the short term. When short-term stores are depleted, the body turns to its long-term stores without problem. Hepatic gluconeogenesis provides the small amount of glucose necessary.

The other persistent myth of starvation mode is that it causes our basal metabolism to decrease severely and our to bodies shut down.

This response, if it were a fact, would also be highly disadvantageous to survival of the human species. If periodic starvation caused our metabolism to decrease, then we would have less energy to hunt or gather food. With less energy, we would be less likely to get food. So, another day passes, and we become even weaker, making us even less likely to get food—a vicious and unsurvivable cycle. It's stupid. There are, in fact, no species of animal, humans included, that have evolved to *require* three meals a day, everyday.

It's unclear to me where this myth originated. Daily caloric restriction does, in fact, lead to decreased metabolism, so people have assumed that this effect would be magnified as food intake dropped to zero. It won't. Decreasing food intake is matched by decreased energy expenditure. However, as food intake goes to zero, the body switches energy inputs from food to stored food (fat). This strategy significantly increases the availability of 'food,' which is matched by an increase in energy expenditure.

So what happened in the Minnesota Starvation Experiment (see chapter 3)? These participants were not fasting, but instead eating a reduced-calorie diet. The hormonal adaptations to fasting were not allowed to happen. Adrenalin was not increased to maintain total energy expenditure. Growth hormone was not increased to maintain lean muscle mass. Ketones were not produced to feed the brain.

Detailed physiologic measurements show that total energy expenditure is *increased* over the duration of a fast.[19] Twenty-two days of alternate daily fasting created no measurable decrease in total energy expenditure. There was no starvation mode. There was no decreased metabolism. Fat oxidation increased 58 per cent, while carbohydrate oxidation decreased from 53 per cent. The body had started to switch over from burning sugar to burning fat, with no overall drop in energy. Four days of continuous fasting actually increased total energy expenditure by 12 per cent.[20] Norepinephrine (adrenalin) levels skyrocketed 117 per cent to maintain energy. Fatty acids increased over 370 per cent as the body switched to burning fat. Insulin decreased 17 per cent. Blood glucose levels dropped slightly, but remained in the normal range.

Concerns are raised repeatedly that fasting may provoke overeating. Studies of caloric intake do show a slight increase at the next meal. After a one-day fast, average caloric intake increased from 2436 to 2914. But over the entire two-day period, there was still a net deficit of 1958 calories. The increased calories on the day after the fast did not nearly make up for the lack of calories on the fasting day.[21] Personal experience in our clinic shows that appetite tends to *decrease* with increased duration of fasting.

FASTING: EXTREME CASES AND GENDER DIFFERENCES

IN 1960, DR. Garfield Duncan of the Pennsylvania Hospital in Philadelphia described his experience with the use of intermittent fasting in the treatment of 107 obese subjects. Subjects who had been unable to lose weight with caloric restriction had lost hope and agreed to try fasting.

One patient (W.H.) started off weighing 325 pounds (147 kilograms) and taking three blood pressure tablets. Over the next fourteen days, he would subsist on nothing but water, tea, coffee and a multivitamin. He found the first two days difficult, but then to his astonishment, his hunger simply vanished. After losing 24 pounds (11 kilograms) in the first fourteen days, he continued with shorter fasting periods, losing a total of 81 pounds (37 kilograms) over the next six months.

Perhaps most surprising was his sense of vigor during the prolonged fasting period.[22] Dr. Duncan wrote, 'A sense of well-being was associated with the fast.'[23] While most expect the fasting period to be extremely difficult, clinicians noted the exact opposite. Dr. E. Drenick wrote, 'The most astonishing aspect of this study was the ease with which prolonged starvation was tolerated.'[24] Others have described the sensation as a mild euphoria[25]—contrasting starkly with the constant hunger, weakness and cold experienced by most low-calorie dieters, as meticulously detailed in the Minnesota Starvation Experiment. These experiences echo our own clinical experience at the Intensive Dietary Management Clinic with hundreds of patients.

Physicians have advocated fasting as far back as the mid 1800s.[26] In modern medicine, reference to fasting can be found as early as 1915,[27] but thereafter it seemed to fall out of favor. In 1951, Dr. W.L. Bloom of Piedmont Hospital in Atlanta 'rediscovered' fasting as a treatment for morbid obesity.[28] Others followed, including Drs. Duncan and Drenick, who described their positive experiences in the *Journal of the American Medical Association.* In an extreme case, in 1973, physicians monitored a man during a 382-day therapeutic fast. Originally weighing 456 pounds, he finished his fast at 180 pounds. No electrolyte abnormalities were noted throughout the period, and the patient felt well throughout.[29]

Several differences are noted in fasting between women and men. Plasma glucose tends to fall faster in women, and ketosis develops more quickly.[30] With increasing body weight, however, the sex difference disappears.[31] Most importantly, the rate of weight loss does not differ substantially between men and women.[32] Personal experience with hundreds of both men and women fail to convince me of any substantial difference between the sexes when it comes to fasting.

INTERMITTENT FASTING AND CALORIC REDUCTION

THE ONE CRUCIAL aspect that differentiates fasting from other diets is its *intermittent* nature. Diets fail because of their constancy. The defining characteristic of life on Earth is homeostasis. Any constant stimulus will eventually be met with an adaptation that resists the change. Persistent exposure to decreased calories results in adaptation (resistance); the body eventually responds by reducing total energy expenditure, leading to the dreaded plateau in weight loss and eventually to weight regain.

A 2011 study compared a portion-control strategy to an intermittent-fasting strategy.[33] The portion-control group reduced daily calories by 25 per cent. For example, if a person normally ate 2000 calories per day, he or she would reduce intake to 1500 calories per day. Over a week, he or she would receive a total of 10,500 calories of a Mediterranean-style diet, which is generally acknowledged to be healthy. The intermittent-fasting group got 100 per cent of their calories for five days of the week, but on the other two days, got only 25 per cent. For example, they received 2000 calories for five days of the week, but on the other two days they would receive only 500—a structure very similar to the 5:2 diet championed by Dr. Michael Mosley. Over a week, they would receive 11,000 calories, slightly more than the portion control group.

At six months, weight loss was similar between the groups (14.3 pounds, or 6.5 kilograms)—but as we know, in the short term, all diets work. However, the intermittent fasting group showed significantly lower insulin levels and insulin resistance. Intermittent diets produced far greater benefits by introducing periods of very low insulin levels that help break the resistance. Further studies confirm that the combination of intermittent fasting with caloric restriction is effective for weight loss.[34,35] The more dangerous visceral fat seems to be preferentially removed. Important risk factors, including LDL cholesterol (low-density lipoproteins), size of low-density lipoproteins and triglycerides, were also improved.

The reverse is also true. Does increasing meal size or frequency contribute to obesity? A recent randomized controlled trial comparing the two demonstrated that only the group with increased eating frequency significantly increased intrahepatic fat.[36] Fatty liver is instrumental in the development of insulin resistance. Increasing the timing of meals has a far more detrimental long-term effect on weight gain. Yet, while we obsess over the question of what to eat, we virtually ignore the crucial aspect of meal timing.

Weight gain is not a steady process. Average yearly weight gain in North Americans is about 1.3 pounds (0.6 kilograms), but that increase is not constant. The year-end holiday period produces a whopping 60 per cent of this yearly weight gain in just six weeks.[37] There is a small weight loss after the holidays, which is not sufficient to counter the gain. In other words, feasting must be followed by fasting. When we remove the fasting and keep all the feasting, we get weight gain.

This is the ancient secret. This is the cycle of life. Fasting follows feasting. Feasting follows fasting. Diets must be *intermittent,* not steady. Food is a celebration of life. Every single culture in the world celebrates with large feasts. That's normal, and it's good. However, religion has always reminded us that we must balance our feasting with periods of fasting—'atonement,' 'repentance' or 'cleansing.' These ideas are ancient and time-tested. Should you eat lots of food on your birthday? Absolutely. Should you eat lots of food at a wedding? Absolutely. These are times to celebrate and indulge. But there is also a time to fast. We cannot change this cycle of life. We cannot feast all the time. We cannot fast all the time. It won't work. It doesn't work.

CAN YOU DO IT?

THOSE WHO HAVE never attempted fasting may be daunted by it. However, as with everything else, fasting becomes much easier with practice. Let's see. Devout Muslims fast for one month of the year and are supposed to fast two days a week. There are an estimated 1.6 *billion* Muslims in the world. There are an estimated 14 million Mormons who are supposed to fast once a month. There are an estimated 350 million Buddhists in the world, many of whom fast regularly. Almost *one-third of the population of the entire world* is supposed to routinely fast throughout their entire lives. There is no question that it can be done. Furthermore, it is clear that there are no lasting negative side effects to regular fasting. Quite the contrary. It appears to have extraordinary health benefits.

Fasting can be combined with any diet imaginable. It makes no difference whether you don't eat meat, dairy or gluten. You can still fast. Eating grass-fed, organic beef is healthy, but can be prohibitively expensive. Fasting contains no hidden costs, but instead saves you money. Eating only homemade, prepared-from-scratch meals is also undoubtedly healthy, but can often be prohibitively time-consuming in our hectic lives. Fasting comes with no time constraints, but instead saves time. No time is required for shopping, food preparation, eating or cleanup.

Life becomes simpler because you do not need to worry about the next meal. Conceptually, fasting is also very simple. The essential elements of fasting can be explained in two minutes. There are no questions such as 'Can I eat whole wheat?' or 'How many calories in that slice of bread?' or 'How many carbs in that pie?' or even 'Are avocados healthy?' The bottom line is that fasting is something that we *can* do, and that we *should* do. See appendix B for some practical tips on successfully introducing fasting into your lifestyle.

So that answers the two unspoken questions. *Is it unhealthy?* The answer is no. Scientific studies conclude that fasting carries significant health benefits. Metabolism increases, energy increases and blood sugars decrease.

The only remaining question is this. *Can you do it?* I hear this one all the time. Absolutely, 100 per cent yes. In fact, fasting has been a part of human culture since the dawn of our species.

'SKIP A FEW MEALS'

ASK A CHILD how to lose weight, and there's a good chance he or she will answer, 'Skip a few meals.' This suggestion is probably the simplest and most correct answer. Instead, we concoct all sorts of intricate rules:

- Eat six times a day.
- Eat a big breakfast.
- Eat low fat.
- Keep a food diary.
- Count your calories.
- Read food labels.
- Avoid all processed foods.
- Avoid white foods—white sugar, white flour, white rice.
- Eat more fiber.

- Eat more fruits and vegetables.
- Mind your microbiome.
- Eat simple foods.
- Eat protein with every meal.
- Eat raw food.
- Eat organic food.
- Count your Weight Watcher points.
- Count your carbs.
- Increase exercise.
- Do resistance and cardio.
- Measure your metabolism and eat less than that.

The list of intricate rules is virtually endless, with more coming every day. It is mildly ironic that even while following this endless list, we're getting fatter than ever. The simple truth is that weight loss comes down to understanding the hormonal roots of obesity. Insulin is the main driver. Obesity is a hormonal, not a caloric imbalance.

There are not one, but two main considerations for proper food choices:

1. What to eat
2. When to eat

In considering the first question, there are some simple guidelines to follow. Reduce intake of refined grains and sugars, moderate protein consumption and increase natural fats. Maximize protective factors such as fiber and vinegar. Choose only natural, unprocessed foods.

In considering the second question, balance insulin-dominant periods with insulin-deficient periods: balance your feeding and fasting. Eating continuously is a recipe for weight gain. Intermittent fasting is a very effective way to deal with when to eat. In the end, the question is this: If you don't eat, will you lose weight? Yes, of course. So there is no real doubt about its efficacy. It will work.

There are other factors that affect insulin and weight loss such as sleep deprivation and stress (cortisol effect). If these are the major pathways of obesity, they must be directly addressed, not with diet, but with techniques such as proper sleep hygiene, meditation, prayer or massage therapy.

For each of us, there will be some factors that are more important than others. For some, sugars may be the main pathway to obesity. For others, it will be chronic sleep deprivation. For yet others, it will be excessive refined grains. For still others, it will be meal timing. Lowering sugar intake will not be so effective if the underlying problem is chronic sleep disturbances. Similarly, better sleep habits will not help if the problem is excessive sugar intake.

What we have tried to develop here is a framework for understanding the complexity of human obesity. A deep and thorough understanding of the causes of obesity leads to rational and successful treatment. A new hope arises. We can begin to dream again—of a world where type 2 diabetes is eradicated, where metabolic syndrome is abolished. A dream of a thinner, healthier tomorrow.

That world. That vision. That dream. It starts today.

APPENDIX A

SAMPLE 7-DAY MEAL PLAN:
24-HOUR FASTING PROTOCOL

	Monday	Tuesday	Wednesd
Breakfast	FAST DAY Water Coffee	Western omelet Green apple	FAST DAY Water Coffee
Lunch	FAST DAY Water Green tea 1 cup of vegetable broth	Arugula salad with walnuts, slices of pear, goat cheese	FAST DAY Water Green tea 1 cup of chicken broth
Dinner	Herbed chicken Green beans	Asian grilled pork belly Baby bok choy stir-fry	Halibut pan-fri in butter and cocunut oil
Dessert	Mixed berries	None	None

These are only meal suggestions. You do not have to follow this particular template.

Refrain from snacking completely.

hursday	Friday	Saturday	Sunday
ran Buds with ed berries	**FAST DAY** Water Coffee	Two eggs Breakfast sausage/bacon Strawberries	**FAST DAY** Water Coffee
ger chicken ce cups fried tables	**FAST DAY** Water Green tea 1 cup of beef broth	Baby spinach and lentil salad	**FAST DAY** Water Green Tea 1 cup of vegetable broth
an chicken y iflower en salad	Baked catfish Sautéed broccoli with garlic and olive oil	Peppered steak Asparagus	Grilled chicken salad
e	Seasonal fruits	None	Dark chocolate

SAMPLE 7-DAY MEAL PLAN:
36-HOUR FASTING PROTOCOL

	Monday	Tuesday	Wednesd
Breakfast	FAST DAY Water Coffee	1 cup of Greek yogurt with ½ cup of mixed blueberries and raspberries, and 1 tbsp of ground flaxseed	FAST DAY Water Coffee
Lunch	FAST DAY Water Green tea 1 cup of vegetable broth	Caesar salad with grilled chicken	FAST DAY Water Green tea 1 cup of chicken broth
Dinner	FAST DAY Water Green tea	Mixed green vegetables sautéed in olive oil Grilled salmon with horseradish sauce	FAST DAY Water Green tea
Dessert	None	Peanut butter on celery sticks	None

These are only meal suggestions. You do not have to follow this particular template.

Refrain from snacking completely.

hursday	Friday	Saturday	Sunday
gs on le	**FAST DAY** Water Coffee	Steel-cut oatmeal with mixed berries and 1 tbsp of ground flaxseed	**FAST DAY** Water Coffee
ger chicken uce cups -fried etables	**FAST DAY** Water Green tea 1 cup of beef broth	Rib-eye steak Grilled vegetables	**FAST DAY** Water Green Tea 1 cup of vegetable broth
ian chicken ry uliflower en salad	**FAST DAY** Water Green tea	Peppered steak Baby bok choy stir-fry	**FAST DAY** Water Green tea
k chocolate: uare of 70% or her Cocoa	None	2 slices of watermelon	None

APPENDIX B
FASTING: A PRACTICAL GUIDE

FASTING IS DEFINED as the voluntary act of withholding food for a specific period of time. Non-caloric drinks such as water and tea are permitted. An absolute fast refers to the withholding of both food and drink. This may be done for religious purposes, such as during Ramadan in the Muslim tradition, but is not generally recommended for health purposes because of the accompanying dehydration.

Fasting has no standard duration. Fasts can range from twelve hours to three months or more. You can fast once a week or once a month or once a year. Intermittent fasting involves fasting for shorter periods of time on a regular basis. Shorter fasts are generally done more frequently. Some people prefer a daily sixteen-hour fast, which means that they eat all their meals within an eight-hour window. Longer fasts are typically twenty-four to thirty-six hours, done two to three times per week. Prolonged fasting may range from one week to one month.

During a twenty-four-hour fast, you fast from dinner (or lunch or breakfast) the first day until dinner (or lunch or breakfast) the next day. Practically, this means missing breakfast, lunch and snacks on the fasting day and only eating a single meal (dinner). Essentially, you skip two meals as you fast from 7:00 p.m. to 7:00 p.m. the next day.

During a thirty-six-hour fast, you fast from dinner on the first day until breakfast two days later. This means missing breakfast, lunch, dinner and snacks for one entire day. You would be skipping three meals as you fast from 7:00 p.m. the first day to 7:00 a.m. two days later. (See appendix A for sample meal plans and fasting protocols.)

Longer fasting periods produce lower insulin levels, greater weight loss and greater blood sugar reduction in diabetics. In the Intensive Dietary Management Clinic, we will typically use a twenty-four-hour or thirty-six-hour fast two to three times per week. For severe diabetes, patients may fast for one to two weeks, but only under close medical supervision. You may take a general multivitamin if you're concerned about micronutrient deficiency.

What can I take on fasting days?

All calorie-containing foods and beverages are withheld during fasting. However, you must stay well hydrated throughout your fast. Water, both still and sparkling, is always a good choice. Aim to drink two liters of water daily. As a good practice, start every day with eight ounces of cool water to ensure adequate hydration as the day begins. Adding a squeeze of lemon or lime flavors the water. Alternatively, you can add some slices of orange or cucumber to a pitcher of water for an infusion of flavor, and then enjoy the water throughout the day. You can dilute apple-cider vinegar in water and then drink it, which may help with your blood sugars. However, artificial flavors or sweeteners are prohibited. Kool-Aid, Crystal Light or Tang should *not* be added to the water.

All types of tea are excellent, including green, black, oolong and herbal. Teas can often be blended together for variety, and can be enjoyed hot or cold. You can use spices such as cinnamon or nutmeg to add flavor to your tea. Adding a small amount of cream or milk is also acceptable. Sugar, artificial sweeteners or flavors are not allowed. Green tea is an especially good choice here. The catechins in green tea are believed to help suppress appetite.

Coffee, caffeinated or decaffeinated, is also permitted. A small amount of cream or milk is acceptable, although these do contain some calories. Spices such as cinnamon may be added, but not sweeteners, sugar or artificial flavors. On hot days, iced coffee is a great choice. Coffee has many health benefits, as previously detailed.

Homemade bone broth, made from beef, pork, chicken or fish bones, is a good choice for fasting days. Vegetable broth is a suitable alternative, although bone broth contains more nutrients. Adding a good pinch of sea salt to the broth will help you stay hydrated. The other fluids—coffee, tea and water—do not contain sodium, so during longer fasting periods, it is possible to become salt-depleted. Although many fear the added sodium, there is far greater danger in becoming salt depleted. For shorter fasts such as the twenty-four- and thirty-six-hour variety, it probably makes little difference. All vegetables, herbs or spices are great additions to broth, but do not add bouillon cubes, which are full of artificial flavors and monosodium glutamate. Beware of canned broths: they are poor imitations of the homemade kinds. (See page 265 for a bone broth recipe.)

Be careful to break your fast gently. Overeating right after fasting may lead to stomach discomfort. While not serious, it can be quite uncomfortable. Instead, try breaking your fast with a handful of nuts or a small salad to start. This problem tends to be self-correcting.

I get hungry when I fast. What can I do?

This is probably the number one concern of fasters everywhere. People assume they'll be overwhelmed with hunger and unable to control themselves. The truth is that hunger does not persist, but instead comes in waves. If you're experiencing hunger, it will pass. Staying busy during a fast day is often helpful. Fasting during a busy day at work keeps your mind off eating.

As the body becomes accustomed to fasting, it starts to burn its stores of fat, and your hunger will be suppressed. Many people note that as they fast, appetite does not increase, but rather starts to decrease. During longer fasts, many people notice that their hunger completely disappears by the second or third day.

There are also natural products that can help suppress hunger. Here are my top five natural appetite suppressants:

1 *Water*: As mentioned before, start your day with a full glass of cold water. Staying hydrated helps prevent hunger. (Drinking a glass of water prior to a meal may also reduce hunger.) Sparkling mineral water may help for noisy stomachs and cramping.
2 *Green tea*: Full of antioxidants and polyphenols, green tea is a great aid for dieters. The powerful antioxidants may help stimulate metabolism and weight loss.
3 *Cinnamon*: Cinnamon has been shown to slow gastric emptying and may help suppress hunger.[1] It may also help lower blood sugars and therefore is useful in weight loss. Cinnamon may be added to all teas and coffees for a delicious change.
4 *Coffee*: While many assume that caffeine suppresses hunger, studies show that this effect is likely related to antioxidants. Both decaffeinated and regular coffee show greater hunger suppression than caffeine in water.[2] Given its health benefits (see chapter 19), there is no reason to limit coffee intake. The caffeine in coffee may also raise your metabolism further boosting fat burning.
5 *Chia Seeds*: Chia seeds are high in soluble fiber and omega 3 fatty acids. These seeds absorb water and form a gel when soaked in liquid for thirty minutes, which may aid in appetite suppression. They can be eaten dry or made into a gel or pudding.

Can I exercise while fasting?

Absolutely. There is no reason to stop your exercise routine. All types of exercise, including resistance (weights) and cardio, are encouraged. There is a common misperception that eating is necessary to supply 'energy' to the working body. That's not true. The liver supplies energy via gluconeogenesis. During longer fasting periods, the muscles are able to use fatty acids directly for energy.

As your adrenalin levels will be higher, fasting is an *ideal* time to exercise. The rise in growth hormone that comes with fasting may also promote muscle growth. These advantages have led many, especially those within the bodybuilding community, to take a greater interest in deliberately exercising in the fasted state. Diabetics on medication, however, must take special precautions because they may experience low blood sugars during exercise and fasting. (See 'What if I have diabetes?' for recommendations, on page 262.)

Will fasting make me tired?

In our experience at the Intensive Dietary Management Clinic, the opposite is true. Many people find that they have *more* energy during a fast—probably due to increased adrenalin. Basal metabolism does not fall during fasting, but rises instead. You'll find you can perform all the normal activities of daily living. Persistent fatigue is not a normal part of fasting. If you experience excessive fatigue, you should stop fasting immediately and seek medical advice.

Will fasting make me confused or forgetful?

No. You should not experience any decrease in memory or concentration. On the contrary, the ancient Greeks believed that fasting significantly improved cognitive abilities, helping the great thinkers attain more clarity and mental acuity. Over the long term, fasting may actually help improve memory. One theory is that fasting activates a form of cellular cleansing called autophagy that may help prevent age-associated memory loss.

I get dizzy when I fast. What can I do?

Most likely, you're becoming dehydrated. Preventing this requires both salt and water. Be sure to drink plenty of fluids. However, the low salt intake on fasting days may cause some dizziness. Extra sea salt in broth or mineral water often helps alleviate the dizziness.

Another possibility is that your blood pressure is too low—particularly if you're taking medications for hypertension. Speak to your physician about adjusting your medications.

I get muscle cramps. What can I do?

Low magnesium levels, particularly common in diabetics, may cause muscle cramps. You may take an over-the-counter magnesium supplement. You may also soak in Epsom salts, which are magnesium salts. Add a cup to a warm bath and soak in it for half an hour. The magnesium will be absorbed through your skin.

I get headaches when I fast. What can I do?

As above, try increasing your salt intake. Headaches are quite common the first few times you try a fast. It is believed that they're caused by the transition from a relatively high-salt diet to very low salt intake on fasting days. Headaches are usually temporary, and as you become accustomed to fasting, this problem often resolves itself. In the meantime, take some extra salt in the form of broth or mineral water.

My stomach is always growling. What can I do?

Try drinking some mineral water.

Since I've started fasting, I experience constipation. What can I do?

Increasing your intake of fiber, fruits and vegetables during the non-fasting period may help with constipation. Metamucil can also be taken to increase fiber and stool bulk. If this problem continues, ask your doctor to consider prescribing a laxative.

I get heartburn. What can I do?

Avoid taking large meals. You may find you have a tendency to overeat once you finish a fast, but try to just eat normally. Breaking a fast is best done slowly. Avoid lying down immediately after a meal and try to stay in an upright position for at least half an hour after meals. Placing wooden blocks under the head of your bed to raise it may help with night-time symptoms. If none of these options work for you, consult your physician.

I take medications with food. What can I do during fasting?

Certain medications may cause problems on an empty stomach. Aspirin can cause stomach upset or even ulcers. Iron supplements may cause nausea and vomiting. Metformin, used for diabetes, may cause nausea or diarrhea. Please discuss whether or not these medications need to be continued with your physician. Also, you can try taking your medications with a small serving of leafy greens.

Blood pressure can sometimes become low during a fast. If you take blood-pressure medications, you may find your blood pressure becomes too low, which can cause light-headedness. Consult with your physician about adjusting your medications.

What if I have diabetes?

Special care must be taken if you are diabetic or are taking diabetic medications. (Certain diabetic medications, such as metformin, are used for other conditions such as polycystic ovarian syndrome.) Monitor your blood sugars closely and adjust your medications accordingly. *Close medical follow-up by your physician is mandatory. If you cannot be followed closely, do not fast.*

Fasting reduces blood sugars. If you are taking diabetic medications, or especially insulin, your blood sugars may become extremely low, which can be a life-threatening situation. You *must* take some sugar or juice to bring your sugars back to normal, even if it means you must stop your fast for that day. *Close monitoring of your blood sugars is mandatory.*

Low blood sugar is *expected* during fasting, so your dose of diabetic medication or insulin may need to be reduced. If you have repeated low blood sugars, it means that you are over-medicated, not that the fasting process is not working. In the Intensive Dietary Management Program, we often reduce medications before starting a fast in anticipation of lower blood sugars. Since the blood sugar response is unpredictable, close monitoring with a physician is essential.

Monitoring

Close monitoring is essential for all patients, but especially, for diabetics. You should also monitor your blood pressure regularly, preferably weekly. Be sure to discuss routine blood work, including electrolyte measurement, with your physician. Should you feel unwell for any reason, stop your fast immediately and seek medical advice. In addition, diabetics should monitor their blood sugars a minimum of twice daily and record this information.

In particular, persistent nausea, vomiting, dizziness, fatigue, high or low blood sugars, or lethargy are not normal with intermittent or continuous fasting. Hunger and constipation are normal symptoms and can be managed.

Intermittent fasting tips

1. *Drink water*: Start each morning with a full eight-ounce glass of water.
2. *Stay busy*: It'll keep your mind off food. It often helps to choose a busy day at work for a fast day.
3. *Drink coffee*: Coffee is a mild appetite suppressant. Green tea, black tea and bone broth may also help.
4. *Ride the waves*: Hunger comes in wave; it is not continuous. When it hits, slowly drink a glass of water or a hot cup of coffee. Often by the time you've finished, your hunger will have passed.
5. *Don't tell everybody you are fasting*: Most people will try to discourage you, as they do not understand the benefits. A close-knit support group is beneficial, but telling everybody you know is not a good idea.
6. *Give yourself one month*: It takes time for your body to get used to fasting. The first few times you fast may be difficult, so be prepared. Don't be discouraged. It will get easier.

7 *Follow a nutritious diet on non-fast days*: Intermittent fasting is not an excuse to eat whatever you like. During non-fasting days, stick to a nutritious diet low in sugars and refined carbohydrates.
8 *Don't binge*: After fasting, pretend it never happened. Eat normally, as if you had never fasted.

The last and most important tip is to fit fasting into your own life! Do not limit yourself socially because you're fasting. Arrange your fasting schedule so that it fits in with your lifestyle. There will be times during which it's impossible to fast: vacation, holidays, weddings. Do not try to force fasting into these celebrations. These occasions are times to relax and enjoy. Afterwards, however, you can simply increase your fasting to compensate. Or just resume your regular fasting schedule. Adjust your fasting schedule to what makes sense for your lifestyle.

What to expect

The amount of weight lost varies tremendously from person to person. The longer that you have struggled with obesity, the more difficult you'll find it to lose weight. Certain medications may make it hard to lose weight. You must simply persist and be patient.

You'll probably eventually experience a weight-loss plateau. Changing either your fasting or dietary regimen, or both, may help. Some patients increase fasting from twenty-four-hour periods to thirty-six-hour periods, or try a forty-eight-hour fast. Some may try eating only once a day, every day. Others may try a continuous fast for an entire week. Changing the fasting protocol is often what's required to break through a plateau.

Fasting is no different than any other skill in life. Practice and support are essential to performing it well. Although it has been a part of human culture forever, many people in North America have never fasted in their lives. Therefore, fasting has been feared and rejected by mainstream nutritional authorities as difficult and dangerous. The truth, in fact, is radically different.

Bone Broth Recipe

Vegetables
Chicken, pork or beef bones
1 tbsp of vinegar
Sea salt, to taste
Pepper, to taste
Ginger root, to taste

1. Add water to cover
2. Simmer for two to three hours until ready
3. Strain and de-fat

APPENDIX C
MEDITATION AND SLEEP HYGIENE TO REDUCE CORTISOL

AS DISCUSSED IN detail in chapter 8, cortisol raises insulin levels and is a major pathway of weight gain. Therefore, reducing your cortisol levels is an integral part of your overall weight-loss effort. Reducing stress levels, practicing meditation and getting good sleep are all effective methods for achieving lower cortisol levels. Some useful tips follow.

Stress reduction

If excessive stress and the cortisol response are causing obesity, then the treatment is to reduce stress, but that's easier said than done. Removing yourself from stressful situations is important, but not always possible. Work and family demands won't go away by themselves. Luckily, there are some time-tested methods of stress relief that can help us cope.

It's a popular misconception that stress relief involves sitting in front of the television and doing nothing. In fact, you can't relieve stress by doing nothing. Stress relief is an active process. Meditation, tai chi, yoga, religious practice and massage are all good choices.

Regular exercise is an excellent way to relieve stress and lower cortisol levels. The original intent of the fight-or-flight response was to mobilize the body for physical exertion. Exercise can also release endorphins and improve mood. This benefit far exceeds the relatively modest caloric reduction achieved by exercise.

Social connectivity is another great stress reliever. Everybody remembers how hard it was to be singled out in high school; that's no different at any age. Being part of a group or community is part of our human heritage. For some, religion and churches can provide this feeling of belonging. The power of human touch also cannot be underestimated. Massage can be beneficial for this reason.

Mindfulness meditation

Through mindfulness meditation, we can become more aware of our thoughts. The objective in meditation is to take a step outside of our thoughts and, as an observer, become aware of them. From this perspective, we can pay precise, nonjudgmental attention to the details of our experiences. Mindfulness meditation alleviates stress by helping us practice being present. It also involves reminding us of pleasant experiences from our past, when we have been able to overcome struggle and achieve personal success. There are many forms of meditation, but all have the same general goals. (Tai chi and yoga are forms of moving meditation with long traditions.)

We don't want to get rid of our thoughts, only become aware of them. We aren't trying to change ourselves, but instead become aware of ourselves as we presently are and objectively observe our thoughts, good or bad.

Meditation can help us work through the thoughts, enabling us to cope with stress much more effectively. Mindfulness meditation can be particularly helpful in working through our feelings of hunger and cravings for foods. Meditation often only takes twenty to thirty minutes and can be done any time. Cultivate the habit of waking up in the morning, having a class of cold water and beginning your meditation.

Three basic aspects are involved in mindfulness meditation: body, breath and thoughts.

Body

First, you want to connect with your body. Find a quiet location where you will not be disturbed for the next twenty minutes. Sit down either on the ground, on a cushion or in a chair. Cross your legs if you are sitting on the ground or on a cushion. If you are sitting on a chair, make sure your feet are placed comfortably on the ground, or on a pillow if your feet do not touch the ground below. It is important that you feel comfortable and relaxed in the position you chose.

Rest your hands on your thighs, palms facing down. Gaze down at the floor about six feet ahead of you and focus on the tip of your nose, and then gently close your eyes. Feel your chest becoming open and your back becoming strong.

Begin your meditation sitting in this position. For a couple of minutes, focus on how your body and your environment feel. If your thoughts wander away from your body, gently bring them back to your body and environment. Do this throughout your meditation every time your mind wanders away.

Breath

Once you have begun to relax, start to focus in gently on your breath. Breathe in through your nose to the count of six and exhale through your mouth slowly to the count of six. Pay attention to how your breath feels entering and exiting your body.

Thoughts

As you sit, you may become bombarded by thoughts. Pay attention to these thoughts. If they cause you to experience any negative emotions, try to think back to a time where you experienced similar challenges and remember how it felt to overcome those challenges. Work though these thoughts until your body begins to feel lighter.

19. Chris Gentilvisio. The 50 Worst Inventions. Time Magazine [Internet]. Available at: http://content.time.com/time/specials/packages/article/0,28804,1991915_1991909_1991785,00.html. Accessed 2015 Apr 15.

Chapter 4: The Exercise Myth

1. British Heart Foundation. Physical activity statistics 2012. Health Promotion Research Group Department of public health, University of Oxford. 2012 Jul. Available from: https://www.bhf.org.uk/~/media/files/research/heart-statistics/m130-bhf_physical-activity-supplement_2012.pdf. Accessed 2015 Apr 8.
2. Public Health England [Internet]. Source data: OEDC. Trends in obesity prevalence. Available from: http://www.noo.org.uk/NOO_about_obesity/trends. Accessed 2015 Apr 8.
3. Countries that exercise the most include United States, Spain, and France. Huffington Post [Internet]. 31 Dec 2013. Available from: http://www.huffingtonpost.ca/2013/12/31/country-exercise-most-_n_4523537.html. Accessed 2015 Apr 6.
4. Dwyer-Lindgren L, Freedman G, Engell RE, Fleming TD, Lim SS, Murray CJ, Mokdad AH. Prevalence of physical activity and obesity in US counties, 2001–2011: a road map for action. Population Health Metrics. 2013 Jul 10; 11:7. Available from http://www.biomedcentral.com/content/pdf/1478-7954-11-7.pdf. Accessed 2015 Apr 8.
5. Byun W, Liu J, Pate RR. Association between objectively measured sedentary behavior and body mass index in preschool children. Int J Obes (Lond). 2013 Jul; 37(7):961–5.
6. Pontzer H. Debunking the hunter-gatherer workout. New York Times [Internet]. 2012 Aug 24. Available from: http://www.nytimes.com/2012/08/26/opinion/sunday/debunking-the-hunter-gatherer-workout.html?_r=0. Accessed 2015 Apr 8.
7. Westerterp KR, Speakman JR. Physical activity energy expenditure has not declined since the 1980s and matches energy expenditure of wild mammals. Int J Obes (Lond). 2008 Aug; 32(8):1256–63.
8. Ross R, Janssen I. Physical activity, total and regional obesity: dose-response considerations. Med Sci Sports Exerc. 2001 Jun; 33(6 Suppl):S521–527.
9. Church TS, Martin CK, Thompson AM, Earnest CP, Mikus CR et al. Changes in weight, waist circumference and compensatory responses with different doses of exercise among sedentary, overweight postmenopausal women. PLoS ONE. 2009; 4(2):e4515. doi:10.1371/journal.pone.0004515. Accessed 2015 Apr 6.
10. Donnelly JE, Honas JJ, Smith BK, Mayo MS, Gibson CA, Sullivan DK, Lee J, Herrmann SD, Lambourne K, Washburn RA. Aerobic exercise alone results in clinically

significant weight loss: Midwest Exercise trial 2. Obesity (Silver Spring). PubMed. 2013 Mar; 21(3):E219–28. doi: 10.1002/oby.20145. Accessed 2015 Apr 6.

11. Church TS et al. Changes in weight, waist circumference and compensatory responses with different doses of exercise among sedentary, overweight postmenopausal women. PLoS ONE. 2009; 4(2):e4515. doi:10.1371/journal.pone.0004515. Accessed 2015 Apr 6.
12. McTiernan A et al. Exercise effect on weight and body fat in men and women. Obesity. 2007 Jun; 15(6):1496–512.
13. Janssen GM, Graef CJ, Saris WH. Food intake and body composition in novice athletes during a training period to run a marathon. Intr J Sports Med. 1989 May; 10(1 suppl.):S17–21.
14. Buring et al. Physical activity and weight gain prevention, Women's Health Study. JAMA. 2010 Mar 24; 303(12):1173–9.
15. Sonneville KR, Gortmaker SL. Total energy intake, adolescent discretionary behaviors and the energy gap. Int J Obes (Lond). 2008 Dec; 32 Suppl 6:S19–27.
16. Child obesity will NOT be solved by PE classes in schools, say researchers. Daily Mail UK [Internet]. 2009 May 7; Health. Available from: http://www.dailymail.co.uk/health/article-1178232/Child-obesity-NOT-solved-PE-classes-schools-say-researchers.html. Accessed 2015 Apr 8.
17. Williams PT, Thompson PD. Increased cardiovascular disease mortality associated with excessive exercise in heart attack survivors. Mayo Clinic Proceedings [Internet]. 2014 Aug. Available from: http://www.mayoclinicproceedings.org/article/S0025-6196%2814%2900437-6/fulltext. DOI: http://dx.doi.org/10.1016/j.mayocp.2014.05.006. Accessed 2015 Apr 8.

Chapter 5: The Overfeeding Paradox

1. Sims EA. Experimental obesity in man. J Clin Invest. 1971 May; 50(5):1005–11.
2. Sims EA et al. Endocrine and metabolic effects of experimental obesity in man. Recent Prog Horm Res. 1973; 29:457–96.
3. Ruppel Shell E. The hungry gene: the inside story of the obesity industry. New York: Grove Press; 2003.
4. Kolata G. Rethinking thin: the new science of weight loss—and the myths and realities of dieting. New York: Farrar, Straus and Giroux; 2008.
5. Levine JA, Eberhardt NL, Jensen MD. Role of nonexercise activity thermogenesis in resistance to fat gain in humans. Science. 1999 Jan 8; 283(5399): 212–4.

6. Diaz EO. Metabolic response to experimental overfeeding in lean and overweight healthy volunteers. Am J Clin Nutr. 1992 Oct; 56(4):641–55.
7. Kechagias S, Ernersson A, Dahlqvist O, Lundberg P, Lindström T, Nystrom FH. Fast-food-based hyper-alimentation can induce rapid and profound elevation of serum alanine aminotransferase in healthy subjects. Gut. 2008 May; 57(5):649–54.
8. DeLany JP, Kelley DE, Hames KC, Jakicic JM, Goodpaster BH. High energy expenditure masks low physical activity in obesity. Int J Obes (Lond). 2013 Jul; 37(7):1006–11.
9. Keesey R, Corbett S. Metabolic defense of the body weight set-point. Res Publ Assoc Res Nerv Ment Dis. 1984; 62:87-96.
10. Leibel RL et al. Changes in energy expenditure resulting from altered body weight. N Engl J Med. 1995 Mar 9; 332(10);621–8.
11. Lustig R. Hypothalamic obesity: causes, consequences, treatment. Pediatr Endocrinol Rev. 2008 Dec; 6(2):220–7.
12. Hervey GR. The effects of lesions in the hypothalamus in parabiotic rat. J Physiol. 1959 Mar 3; 145(2):336–52.3.
13. Heymsfield SB et al. Leptin for weight loss in obese and lean adults: a randomized, controlled, dose-escalation trial. JAMA. 1999 Oct 27; 282(16):1568–75.

Chapter 6: A New Hope

1. Tentolouris N, Pavlatos S, Kokkinos A, Perrea D, Pagoni S, Katsilambros N. Diet-induced thermogenesis and substrate oxidation are not different between lean and obese women after two different isocaloric meals, one rich in protein and one rich in fat. Metabolism. 2008 Mar; 57(3):313–20.
2. Data source for Figure 6.1: Ibid.

Chapter 7: Insulin

1. Polonski K, Given B, Van Cauter E. Twenty-four hour profiles and pulsatile patterns of insulin secretion in normal and obese subjects. J Clin Invest. 1988 Feb; 81(2):442–8.
2. Ferrannini E, Natali A, Bell P, et al. Insulin resistance and hypersecretion in obesity. J Clin Invest. 1997 Sep 1; 100(5):1166–73.
3. Han TS, Williams K, Sattar N, Hunt KJ, Lean ME, Haffner SM. Analysis of obesity and hyperinsulinemia in the development of metabolic syndrome: San Antonio Heart Study. Obes Res. 2002 Sep; 10(9):923–31.

4. Russell-Jones D, Khan R. Insulin-associated weight gain in diabetes: causes, effects and coping strategies. Diabetes, Obesity and Metabolism. 2007 Nov; 9(6):799–812.

5. White NH et al. Influence of intensive diabetes treatment on body weight and composition of adults with type 1 diabetes in the Diabetes Control and Complications Trial. Diabetes Care. 2001; 24(10):1711–21.

6. Intensive blood-glucose control with sulphonylureas or insulin compared with conventional treatment and risk of complications in patients with type 2 diabetes (UKPDS33). Lancet. 1998 Sep 12; 352(9131):837–53.

7. Holman RR et al. Addition of biphasic, prandial, or basal insulin to oral therapy in type 2 diabetes. N Engl J Med. 2007 Oct 25; 357(17):1716–30.

8. Henry RR, Gumbiner B, Ditzler T, Wallace P, Lyon R, Glauber HS. Intensive conventional insulin therapy for type II diabetes. Diabetes Care. 1993 Jan; 16(1):23–31.

9. Doherty GM, Doppman JL, Shawker TH, Miller DL, Eastman RC, Gorden P, Norton JA. Results of a prospective strategy to diagnose, localize, and resect insulinomas. Surgery. 1991 Dec; 110(6):989–96.

10. Ravnik-Oblak M, Janez A, Kocijanicic A. Insulinoma induced hypoglycemia in a type 2 diabetic patient. Wien KlinWochenschr. 2001 Apr 30; 113(9):339–41.

11. Sapountzi P et al. Case study: diagnosis of insulinoma using continuous glucose monitoring system in a patient with diabetes. Clin Diab. 2005 Jul; 23(3):140–3.

12. Smith CJ, Fisher M, McKay GA. Drugs for diabetes: part 2 sulphonylureas. Br J Cardiol. 2010 Nov; 17(6):279–82.

13. Viollet B, Guigas B, Sanz Garcia N, Leclerc J, Foretz M, Andreelli F. Cellular and molecular mechanisms of metformin: an overview. Clin Sci (Lond). 2012 Mar; 122(6):253–70.

14. Klip A, Leiter LA. Cellular mechanism of action of metformin. Diabetes Care. 1990 Jun; 13(6):696–704.

15. King P, Peacock I, Donnelly R. The UK Prospective Diabetes Study (UKPDS): clinical and therapeutic implications for type 2 diabetes. Br J Clin Pharmacol. 1999 Nov; 48(5):643–8.

16. UK Prospective Diabetes Study (UKPDS) Group. Effect of intensive blood-glucose control with metformin on complications in overweight patients with type 2 diabetes (UKPDS34). Lancet. 1998 Sep 12; 352(9131):854–65.

17. DeFronzo RA, Ratner RE, Han J, Kim DD, Fineman MS, Baron AD. Effects of exenatide (exendin-4) on glycemic control and weight over 30 weeks in metformin-treated patients with type 2 diabetes. Diabetes Care. 2004 Nov; 27(11):2628–35.

18. Nauck MA, Meininger G, Sheng D, Terranella L, Stein PP. Efficacy and safety of the dipeptidyl peptidase-4 inhibitor, sitagliptin, compared with the sulfonylurea, glipizide, in patients with type 2 diabetes inadequately controlled on metformin alone: a randomized, double-blind, non-inferiority trial. Diabetes Obes Metab. 2007 Mar; 9(2): 194–205.

19. Meneilly GS et al. Effect of acarbose on insulin sensitivity in elderly patients with diabetes. Diabetes Care. 2000 Aug; 23(8):1162–7.

20. Wolever TM, Chiasson JL, Josse RG, Hunt JA, Palmason C, Rodger NW, Ross SA, Ryan EA, Tan MH. Small weight loss on long-term acarbose therapy with no change in dietary pattern or nutrient intake of individuals with non-insulin-dependent diabetes. Int J Obes Relat Metab Disord. 1997 Sep; 21(9):756–63.

21. Polidori D et al. Canagliflozin lowers postprandial glucose and insulin by delaying intestinal glucose absorption in addition to increasing urinary glucose excretion: results of a randomized, placebo-controlled study. Diabetes Care. 2013 Aug; 36(8):2154–6.

22. Bolinder J et al. Effects of dapagliflozin on body weight, total fat mass, and regional adipose tissue distribution in patients with type 2 diabetes mellitus with inadequate glycemic control on metformin. J Clin Endocrinol Metab. 2012 Mar; 97(3):1020–31.

23. Nuack MA et al. Dapagliflozin versus glipizide as add-on therapy in patients with type 2 diabetes who have inadequate glycemic control with metformin. Diabetes Care. 2011 Sep; 34(9):2015–22.

24. Domecq JP et al. Drugs commonly associated with weight change: a systematic review and meta-analysis. J Clin Endocrinol Metab. 2015 Feb; 100(2):363–70.

25. Ebenbichler CF et al. Olanzapine induces insulin resistance: results from a prospective study. J Clin Psychiatry. 2003 Dec; 64(12):1436–9.

26. Scholl JH, van Eekeren, van Puijenbroek EP. Six cases of (severe) hypoglycaemia associated with gabapentin use in both diabetic and non-diabetic patients. Br J Clin Pharmacol. 2014 Nov 11. doi: 10.1111/bcp.12548. [Epub ahead of print.] Accessed 2015 Apr 6.

27. Penumalee S, Kissner P, Migdal S. Gabapentin induced hypoglycemia in a long-term peritoneal dialysis patient. Am J Kidney Dis. 2003 Dec; 42(6):E3–5.

28. Suzuki Y et al. Quetiapine-induced insulin resistance after switching from blonanserin despite a loss in both bodyweight and waist circumference. Psychiatry Clin Neurosci. 2012 Oct; 66(6):534–5.

29. Kong LC et al. Insulin resistance and inflammation predict kinetic body weight changes in response to dietary weight loss and maintenance in overweight and

obese subjects by using a Bayesian network approach. Am J Clin Nutr. 2013 Dec; 98(6):1385–94.

30. Lustig RH et al. Obesity, leptin resistance, and the effects of insulin suppression. Int J Obesity. 2004 Aug 17; 28:1344–8.
31. Martin SS, Qasim A, Reilly MP. Leptin resistance: a possible interface of inflammation and metabolism in obesity-related cardiovascular disease. J Am Coll Cardiol. 2008 Oct 7; 52(15):1201–10.
32. Benoit SC, Clegg DJ, Seeley RJ, Woods SC. Insulin and leptin as adiposity signals. Recent Prog Horm Res. 2004; 59:267–85.

Chapter 8: Cortisol

1. Owen OE, Cahill GF Jr. Metabolic effects of exogenous glucocorticoids in fasted man. J Clin Invest. 1973 Oct; 52(10):2596–600.
2. Rosmond R et al. Stress-related cortisol secretion in men: relationships with abdominal obesity and endocrine, metabolic and hemodynamic abnormalities. J Clin Endocrinol Metab. 1998 Jun; 83(6):1853–9.
3. Whitworth JA et al. Hyperinsulinemia is not a cause of cortisol-induced hypertension. Am J Hypertens. 1994 Jun; 7(6):562–5.
4. Pagano G et al. An in vivo and in vitro study of the mechanism of prednisone-induced insulin resistance in healthy subjects. J Clin Invest. 1983 Nov; 72(5):1814–20.
5. Rizza RA, Mandarino LJ, Gerich JE. Cortisol-induced insulin resistance in man: impaired suppression of glucose production and stimulation of glucose utilization due to a postreceptor detect of insulin action. J Clin Endocrinol Metab. 1982 Jan; 54(1):131–8.
6. Ferris HA, Kahn CR. New mechanisms of glucocorticoid-induced insulin resistance: make no bones about it. J Clin Invest. 2012 Nov; 122(11):3854–7.
7. Stolk RP et al. Gender differences in the associations between cortisol and insulin in healthy subjects. J Endocrinol. 1996 May; 149(2):313–8.
8. Jindal RM et al. Posttransplant diabetes mellitus: a review. Transplantation. 1994 Dec 27; 58(12):1289–98.
9. Pagano G et al. An in vivo and in vitro study of the mechanism of prednisone-induced insulin resistance in healthy subjects. J Clin Invest. 1983 Nov; 72(5):1814–20.
10. Rizza RA, Mandarino LJ, Gerich JE. Cortisol-induced insulin resistance in man: impaired suppression of glucose production and stimulation of glucose

utilization due to a postreceptor defect of insulin action. J Clin Endocrinol Metab. 1982 Jan; 54(1):131–8.

11. Dinneen S, Alzaid A, Miles J, Rizza R. Metabolic effects of the nocturnal rise in cortisol on carbohydrate metabolism in normal humans. J Clin Invest. 1993 Nov; 92(5):2283–90.
12. Lemieux I et al. Effects of prednisone withdrawal on the new metabolic triad in cyclosporine-treated kidney transplant patients. Kidney International. 2002 Nov; 62(5):1839–47.
13. Fauci A et al., editors. Harrison's principles of internal medicine. 17th ed. McGraw-Hill Professional; 2008. p. 2255.
14. Tauchmanova L et al. Patients with subclinical Cushing's syndrome due to adrenal adenoma have increased cardiovascular risk. J Clin Endocrinol Metab. 2002 Nov; 87(11):4872–8.
15. Fraser R et al. Cortisol effects on body mass, blood pressure, and cholesterol in the general population. Hypertension. 1999 Jun; 33(6):1364–8.
16. Marin P et al. Cortisol secretion in relation to body fat distribution in obese premenopausal women. Metabolism. 1992 Aug; 41(8):882–6.
17. Wallerius S et al. Rise in morning saliva cortisol is associated with abdominal obesity in men: a preliminary report. J Endocrinol Invest. 2003 Jul; 26(7):616–9.
18. Wester VL et al. Long-term cortisol levels measured in scalp hair of obese patients. Obesity (Silver Spring). 2014 Sep; 22(9):1956–8. DOI: 10.1002/oby.20795. Accessed 2015 Apr 6.
19. Fauci A et al., editors. Harrison's principles of internal medicine. 17th ed. McGraw-Hill Professional; 2008. p. 2263.
20. Daubenmier J et al. Mindfulness intervention for stress eating to reduce cortisol and abdominal fat among overweight and obese women. Journal of Obesity. 2011; article ID 651936. Accessed 2015 Apr 6.
21. Knutson KL, Spiegel K, Penev P, van Cauter E. The metabolic consequences of sleep deprivation. Sleep Med Rev. 2007 Jun; 11(3):163–78.
22. Webb WB, Agnew HW. Are we chronically sleep deprived? Bull Psychon Soc. 1975; 6(1):47–8.
23. Bliwise DL. Historical change in the report of daytime fatigue. Sleep. 1996 Jul; 19(6):462–4.
24. Watanabe M et al. Association of short sleep duration with weight gain and obesity at 1-year follow-up: a large-scale prospective study. Sleep. 2010 Feb; 33(2):161–7.

25. Hasler G, Buysse D, Klaghofer R, Gamma A, Ajdacic V, et al. The association between short sleep duration and obesity in young adults: A 13-year prospective study. Sleep. 2004 Jun 15; 27(4):661–6.
26. Cappuccio FP et al. Meta-analysis of short sleep duration and obesity in children and adults. Sleep. 2008 May; 31(5):619–26.
27. Joo EY et al. Adverse effects of 24 hours of sleep deprivation on cognition and stress hormones. J Clin Neurol. 2012 Jun; 8(2):146–50.
28. Leproult R et al. Sleep loss results in an elevation of cortisol levels the next evening. Sleep. 1997 Oct; 20(10):865–70.
29. Spiegel K, Knutson K, Leproult R, Tasali E, Cauter EV. Sleep loss: a novel risk factor for insulin resistance and Type 2 diabetes. J Appl Physiol. 2005 Nov; 99(5):2008–19.
30. VanHelder T, Symons JD, Radomski MW. Effects of sleep deprivation and exercise on glucose tolerance. Aviat Space Environ Med. 1993 Jun; 64(6):487–92.
31. Sub-chronic sleep restriction causes tissue specific insulin resistance. J Clin Endocrinol Metab. 2015 Feb 6; jc20143911. [Epub ahead of print] Accessed 2015 Apr 6.
32. Kawakami N, Takatsuka N, Shimizu H. Sleep disturbance and onset of type 2 diabetes. Diabetes Care. 2004 Jan; 27(1):282–3.
33. Taheri S, Lin L, Austin D, Young T, Mignot E. Short sleep duration is associated with reduced leptin, elevated ghrelin, and increased body mass index. PLoS Medicine. 2004 Dec; 1(3):e62.
34. Nedeltcheva AV et al. Insufficient sleep undermines dietary efforts to reduce adiposity. Ann Int Med. 2010 Oct 5; 153(7):435–41.
35. Pejovic S et al. Leptin and hunger levels in young healthy adults after one night of sleep loss. J. Sleep Res. 2010 Dec; 19(4):552–8.

Chapter 9: The Atkins Onslaught

1. Pennington AW. A reorientation on obesity. N Engl J Med. 1953 Jun 4; 248(23):959–64.
2. Bloom WL, Azar G, Clark J, MacKay JH. Comparison of metabolic changes in fasting obese and lean patients. Ann NY Acad Sci. 1965 Oct 8; 131(1):623–31.
3. Stillman I. The doctor's quick weight loss diet. Ishi Press; 2011.
4. Kolata G. Rethinking thin: the new science of weight loss—and the myths and realities of dieting. Picador; 2008.
5. Samaha FF et al. A low-carbohydrate as compared with a low-fat diet in severe obesity. N Engl J Med. 2003 May 22; 348(21):2074–81.

6. Gardner CD et al. Comparison of the Atkins, Zone, Ornish, and LEARN diets for change in weight and related risk factors among overweight premenopausal women. JAMA. 2007 Mar 7; 297(9):969–77.
7. Shai I et al. Weight loss with a low-carbohydrate, Mediterranean, or low-fat die. N Engl J Med. 2008 Jul 17; 359(3):229–41.
8. Larsen TM et al. Diets with high or low protein content and glycemic index for weight-loss maintenance. N Engl J Med. 2010 Nov 25; 363(22):2102–13.
9. Ebbeling C et al. Effects of dietary composition on energy expenditure during weight-loss maintenance. JAMA. 2012 Jun 27; 307(24):2627–34.
10. Boden G et al. Effect of a low-carbohydrate diet on appetite, blood glucose levels, and insulin resistance in obese patients with type 2 diabetes. Ann Intern Med. 2005 Mar 15; 142(6):403–11.
11. Foster G et al. Weight and metabolic outcomes after 2 years on a low-carbohydrate versus low-fat diet. Ann Int Med. 2010 Aug 3; 153(3):147–57.
12. Shai I et al. Four-year follow-up after two-year dietary interventions. N Engl J Med. 2012 Oct 4; 367(14):1373–4.
13. Hession M et al. Systematic review of randomized controlled trials of low-carbohydrate vs. low-fat/low calorie diets in the management of obesity and its comorbidities. Obes Rev. 2009 Jan; 10(1):36–50.
14. Zhou BG et al. Nutrient intakes of middle-aged men and women in China, Japan, United Kingdom, and United States in the late 1990s: The INTERMAP Study. J Hum Hypertens. 2003 Sep; 17(9):623–30.
15. Data source for Figure 9.1: Ibid.
16. Lindeberg S et al. Low serum insulin in traditional Pacific Islanders: the Kitava Study. Metabolism. 1999 Oct; 48(10):1216–9.

Chapter 10: Insulin Resistance: The Major Player

1. Tirosh A et al. Adolescent BMI trajectory and risk of diabetes versus coronary disease. N Engl J Med. 2011 Apr 7; 364(14):1315–25.
2. Alexander Fleming. Penicillin. Nobel Lecture Dec 1945. Available from: http://www.nobelprize.org/nobel_prizes/medicine/laureates/1945/fleming-lecture.pdf. Accessed 2015 Apr 15.
3. Pontiroli AE, Alberetto M, Pozza G. Patients with insulinoma show insulin resistance in the absence of arterial hypertension. Diabetologia. 1992 Mar; 35(3):294–5.

4. Pontiroli AE, Alberetto M, Capra F, Pozza G. The glucose clamp technique for the study of patients with hypoglycemia: insulin resistance as a feature of insulinoma. J Endocrinol Invest. 1990 Mar; 13(3):241–5.

5. Ghosh S et al. Clearance of acanthosis nigricans associated with insulinoma following surgical resection. QJM. 2008 Nov; 101(11):899–900. doi: 10.1093/qjmed/hcn098. Epub 2008 Jul 31. Accessed 2015 Apr 8.

6. Rizza RA et al. Production of insulin resistance by hyperinsulinemia in man. Diabetologia. 1985 Feb; 28(2):70–5.

7. Del Prato S et al. Effect of sustained physiologic hyperinsulinemia and hyperglycemia on insulin secretion and insulin sensitivity in man. Diabetologia. 1994 Oct; 37(10):1025–35.

8. Henry RR et al. Intensive conventional insulin therapy for type II diabetes. Diabetes Care. 1993 Jan; 16(1):23–31.

9. Le Stunff C, Bougneres P. Early changes in postprandial insulin secretion, not in insulin sensitivity characterize juvenile obesity. Diabetes. 1994 May; 43(5):696–702.

10. Popkin BM, Duffey KJ. Does hunger and satiety drive eating anymore? Am J Clin Nutr. 2010 May; 91(5):1342–7.

11. Duffey KJ, Popkin BM. Energy density, portion size, and eating occasions: contributions to increased energy intake in the United States, 1977–2006. PLoS Med. 2011 Jun; 8(6): e1001050. doi:10.1371/journal.pmed.1001050. Accessed 2015 Apr 8.

12. Bellisle F, McDevitt R, Prentice AM. Meal frequency and energy balance. Br J Nutr. 1997 Apr; 77 Suppl 1:S57–70.

13. Cameron JD, Cyr MJ, Doucet E. Increased meal frequency does not promote greater weight loss in subjects who were prescribed an 8-week equi-energetic energy-restricted diet. Br J Nutr. 2010 Apr; 103(8):1098–101.

14. Leidy JH et al. The influence of higher protein intake and greater eating frequency on appetite control in overweight and obese men. Obesity (Silver Spring). 2010 Sep; 18(9):1725–32.

15. Stewart WK, Fleming LW. Features of a successful therapeutic fast of 382 days' duration. Postgrad Med J. 1973 Mar; 49(569):203–09.

Chapter 11: Big Food, More Food and the New Science of Diabesity

1. Center for Science in the Public Interest [Internet]. Non-profit organizations receiving corporate funding. Available from: http://www.cspinet.org/integrity/nonprofits/american_heart_association.html. Accessed 2015 Apr 8.

2. Freedhoff, Y. Weighty Matters blog [Internet]. Heart and Stroke Foundation Health Check on 10 teaspoons of sugar in a glass. 2012 Apr 9. Available from: http://www.weightymatters.ca/2012/04/heart-and-stroke-foundation-health.html. Accessed 2015 Apr 8.

3. Lesser LI, Ebbeling CB, Goozner M, Wypij D, Ludwig D. Relationship between funding source and conclusion among nutrition-related scientific articles. PLoS Med. 2007 Jan 9; 4(1): e5. doi:10.1371/journal.pmed.0040005. Accessed 2015 Apr 8.

4. Nestle M. Food company sponsorship of nutrition research and professional activities: A conflict of interest? Public Health Nutr. 2001 Oct; 4(5):1015–22.

5. Stubbs RJ, Mazlan N, Whybrow S. Carbohydrates, appetite and feeding behavior in humans. J Nutr. 2001 Oct 1; 131(10):2775–81S.

6. Cameron JD, Cyr MJ, Doucet E. Increased meal frequency does not promote greater weight loss in subjects who were prescribed an 8-week equi-energetic energy-restricted diet. Br J Nutr. 2010 Apr; 103(8):1098–101.

7. Wyatt HR et al. Long-term weight loss and breakfast in subjects in the National Weight Control Registry. Obes Res. 2002 Feb; 10(2):78–82.

8. Wing RR, Phelan S. Long term weight loss maintenance. Am J Clin Nutr. 2005 Jul; 82(1 Suppl):222S–5S.

9. Brown AW et al. Belief beyond the evidence: using the proposed effect of breakfast on obesity to show 2 practices that distort scientific evidence. Am J Clin Nutr. 2013 Nov; 98(5):1298–308.

10. Schusdziarra V et al. Impact of breakfast on daily energy intake. Nutr J. 2011 Jan 17; 10:5. doi: 10.1186/1475-2891-10-5. Accessed 2015 Apr 8.

11. Reeves S et al. Experimental manipulation of breakfast in normal and overweight/obese participants is associated with changes to nutrient and energy intake consumption patterns. Physiol Behav. 2014 Jun 22; 133:130–5. doi: 10.1016/j.physbeh.2014.05.015. Accessed 2015 Apr 8.

12. Dhurandhar E et al. The effectiveness of breakfast recommendations on weight loss: a randomized controlled trial. Am J Clin Nutr. 2014 Jun 4. doi: 10.3945/ajcn.114.089573. Accessed 2015 Apr 8.

13. Betts JA et al. The causal role of breakfast in energy balance and health: a randomized controlled trial in lean adults. Am J Clin Nutr. 2014 Aug; 100(2): 539–47.

14. Diet, nutrition and the prevention of chronic disease: report of a joint WHO/FAO expert consultation. Geneva: World Health Organization; 2003. p. 68. Available at: http://whqlibdoc.who.int/trs/who_trs_916.pdf. Accessed 2015 Apr 9.

15. Kaiser KA et al. Increased fruit and vegetable intake has no discernible effect on weight loss: a systematic review and meta-analysis. Am J Clin Nutr. 2014 Aug; 100(2):567–76.

16. Muraki I et al. Fruit consumption and the risk of type 2 Diabetes. BMJ. 2013 Aug 28; 347:f5001. doi: 10.1136/bmj.f5001. Accessed 2015 Apr 8.

Chapter 12: Poverty and Obesity

1. Centers for Disease Control and Prevention. Obesity trends among U.S. adults between 1985 and 2010. Available from: www.cdc.gov/obesity/downloads/obesity_trends_2010.ppt. Accessed 2015 Apr 26.

2. United States Census Bureau [Internet]. State and country quick facts. Updated 2015 Mar 24. Available from: http://quickfacts.census.gov/qfd/states/28000.html. Accessed 2015 Apr 8.

3. Levy J. Mississippians most obese, Montanans least obese. Gallup [Internet]. Available from: http://www.gallup.com/poll/167642/mississippians-obese-montanans-least-obese.aspx. Accessed 2015 Apr 8.

4. Michael Moss. Salt Sugar Fat: How the Food Giants Hooked Us. Toronto; Signal Publishing; 2014.

5. David Kessler. The End of Overeating: Taking Control of the Insatiable North American Appetite. Toronto: McClelland & Stewart Publishing; 2010.

6. Data source for Figure 12.2: Environmental Working Group (EWG). EWG farm subsidies. Available from: http://farm.ewg.org/. Accessed 2015 Apr 26.

7. Russo M. Apples to twinkies: comparing federal subsidies of fresh produce and junk food. US PIRG Education Fund: 2011 Sep. Available at: http://www.foodsafetynews.com/files/2011/09/Apples-to-Twinkies-USPIRG.pdf. Accessed 2015 Apr 26.

8. Data source for Figure 12.3: Ibid.

9. Mills CA: Diabetes mellitus: is climate a responsible factor in the etiology? Arch Inten Med. 1930 Oct; 46(4):569–81.

10. Marchand LH. The Pima Indians: Obesity and diabetes. National Diabetes Information Clearinghouse (NDICH) [Internet]. Available from: https://web.archive.org/web/20150610193111. Accessed 2015 Apr 8.

11. U.S. PIRG [Internet].Report: 21st century transportation. 2013 May 14. Available from: http://uspirg.org/reports/usp/new-direction. Accessed 2015 Apr 8.

12. Davies A. The age of the car in America is over. Business Insider [Internet]. 2013 May 20. http://www.businessinsider.com/the-us-driving-boom-is-over-2013-5. Accessed 2015 Apr 8.

Chapter 13: Childhood Obesity

1. Foster GD et al. The HEALTHY Study Group. A school-based intervention for diabetes risk reduction. N Engl J Med. 2010 Jul 29; 363(5):443–53.
2. Must A, Jacques PF, Dallal GE, Bajema CJ, Dietz WH. Long-term morbidity and mortality of overweight adolescents: a follow-up of the Harvard Growth Study of 1922 to 1935. N Engl J Med. 1992 Nov; 327(19):1350–5.
3. Deshmukh-Taskar P, Nicklas TA, Morales M, Yang SJ, Zakeri I, Berenson GS. Tracking of overweight status from childhood to young adulthood: the Bogalusa Heart Study. Eur J Clin Nutr. 2006 Jan; 60(1):48–57.
4. Baker JL, Olsen LW, Sørensen TI. Childhood body-mass index and the risk of coronary heart disease in adulthood. N Engl J Med. 2007 Dec; 357(23):2329–37.
5. Juonala M et al. Childhood adiposity, adult adiposity, and cardiovascular risk factors. N Engl J Med. 2011 Nov 17; 365(20):1876–85.
6. Kim J et al. Trends in overweight from 1980 through 2001 among preschool-aged children enrolled in a health maintenance organization. Obesity (Silver Spring). 2006 Jul; 14(7):1107–12.
7. Bergmann RL et al. Secular trends in neonatal macrosomia in Berlin: influences of potential determinants. Paediatr Perinat Epidemiol. 2003 Jul; 17(3):244–9.
8. Holtcamp W. Obesogens: an environmental link to obesity. Environ Health Perspect. 2012 Feb; 120(2):a62–a68.
9. Ludwig DS, Currie J. The association between pregnancy weight gain and birth weight. Lancet. 2010 Sep 18; 376(9745):984–90.
10. Whitaker RC et al. Predicting obesity in young adulthood from childhood and parental obesity. N Engl J Med. 1997 Sep 25; 337(13):869–73.
11. Caballero B et al. Pathways: A school-based randomized controlled trial for the prevention of obesity in American Indian schoolchildren. Am J Clin Nutr. 2003 Nov; 78(5):1030–8.
12. Nader PR et al. Three-year maintenance of improved diet and physical activity: the CATCH cohort. Arch Pediatr Adoles Med. 1999 Jul; 153(7):695-705.
13. Klesges RC et al. The Memphis Girls Health Enrichment Multi-site Studies (GEMS): Arch Pediatr Adolesc Med. 2010 Nov; 164(11):1007–14.
14. de Silva-Sanigorski AM et al. Reducing obesity in early childhood: results from Romp & Chomp, an Australian community-wide intervention program. Am J Clin Nutr. 2010 Apr; 91(4):831–40.

15. James J et al. Preventing childhood obesity by reducing consumption of carbonated drinks: cluster randomised controlled trial. BMJ. 2004 May 22; 328(7450):1237.
16. Ogden CL et al. Prevalence of childhood and adult obesity in the United States, 2011–2012. JAMA.2014 Feb 26; 311(8):806–14.
17. Spock B. Doctor Spock's baby and child care. Pocket Books; 1987. p. 536.

Chapter 14: The Deadly Effects of Fructose

1. Suddath C, Stanford D. Coke confronts its big fat problem. Bloomberg Businessweek [Internet]. 2014 July 31. Available from: http://www.bloomberg.com/bw/articles/2014-07-31/coca-cola-sales-decline-health-concerns-spur-relaunch Accessed 2015 Apr 8.
2. Ibid.
3. S&D (Group sucres et denrées) [Internet]. World sugar consumption. Available from: http://www.sucden.com/statistics/4_world-sugar-consumption. Accessed 2015 Apr 9.
4. Xu Y et al. Prevalence and control of diabetes in Chinese adults. JAMA. 2013 Sep 4; 310(9):948–59.
5. Loo D. China 'catastrophe' hits 114 million as diabetes spreads. Bloomberg News [Internet]. 2013 Sep 3. Available from: http://www.bloomberg.com/news/articles/2013-09-03/china-catastrophe-hits-114-million-as-diabetes-spreads. Accessed 2015 Apr 8.
6. Huang Y. China's looming diabetes epidemic. The Atlantic [Internet]. 2013 Sept 13. Available from: http://www.theatlantic.com/china/archive/2013/09/chinas-looming-diabetes-epidemic/279670/. Accessed 2015 Apr 8.
7. Schulze MB et al. Sugar-sweetened beverages, weight gain and incidence of type 2 diabetes in young and middle aged women. JAMA. 2004 Aug 25; 292(8):927–34.
8. Basu S, Yoffe P, Hills N, Lustig RH. The relationship of sugar to population-level diabetes prevalence: an econometric analysis of repeated cross-sectional data. Plos One [Internet]. 2013; 8(2):e57873 doi: 10.1371/journal.pone.0057873. Accessed 2015 Apr 8.
9. Lyons RD. Study insists diabetics can have some sugar. New York Times [Internet]. 1983 Jul 7. Available from: http://www.nytimes.com/1983/07/07/us/study-insists-diabetics-can-have-some-sugar.html. Accessed 2015 Apr 8.
10. Glinsmann WH et al. Evaluation of health aspects of sugars contained in carbohydrate sweeteners. J Nutr. 1986 Nov; 116(11S):S1–S216.

11. National Research Council (US) Committee on Diet and Health. Diet and health: implications for reducing chronic disease risk. Washington (DC): National Academies Press (US); 1989. p. 7.

12. American Diabetes Association [Internet]. Sugar and desserts. Edited 2015 Jan 27. Available from: http://www.diabetes.org/food-and-fitness/food/what-can-i-eat/understanding-carbohydrates/sugar-and-desserts.html. Accessed 2015 Apr 8.

13. Zhou BF et al. Nutrient intakes of middle-aged men and women in China, Japan, United Kingdom, and United States in the late 1990s. J Hum Hypertens. 2003 Sep; 17(9):623–30.

14. Duffey KJ, Popkin BM. High-Fructose Corn syrup: Is this what's for dinner? Am J Clin Nutr. 2008; 88(suppl):1722S–32S.

15. Bray GA, Nielsen SJ, Popkin BM. Consumption of high-fructose corn syrup in beverages may play a role in the epidemic of obesity. Am J Clin Nutr. 2004 April; 79(4) 537–43.

16. Beck-Nielsen H et al. Impaired cellular insulin binding and insulin sensitivity induced by high-fructose feeding in normal subjects. Am J Clin Nutr. 1980 Feb; 33(2):273–8.

17. Stanhope KL et al. Consuming fructose-sweetened, not glucose-sweetened, beverages increases visceral adiposity and lipids and decreases insulin sensitivity in overweight/obese humans. JCI. 2009 May 1; 119(5):1322–34.

18. Sievenpiper JL et al. Effect of fructose on body weight in controlled feeding trials: a systematic review and meta-analysis. Ann Intern Med. 2012 Feb 21; 156(4):291–304.

19. Ogden CL et al. Prevalence of childhood and adult obesity in the United States, 2011–2012. JAMA. 2014 Feb 26; 311(8):806–14.

20. Geiss LS et al. Prevalence and incidence trends for diagnosed diabetes among adults aged 20 to 79 years, United States, 1980–2012. JAMA. 2014 Sep 24; 312(12):1218–26.

Chapter 15: The Diet Soda Delusion

1. Yang Q. Gain weight by 'going diet?' Artificial sweeteners and the neurobiology of sugar cravings. Yale J Biol Med. 2010 Jun; 83(2):101–8.

2. Mattes RD, Popkin BM. Nonnutritive sweetener consumption in humans: effects on appetite and food intake and their putative mechanisms. Am J Clin Nutr. 2009 Jan; 89(1):1–14. (This article is also the data source for Figure 15.1.)

3. Gardner C et al. Nonnutritive sweeteners: current use and health perspectives: a scientific statement from the American Heart Association and the American Diabetes Association. Circulation. 2012 Jul 24; 126(4):509–19.

4. Oz, M. Agave: why we were wrong. The Oz Blog. 2014 Feb 27. Available from: http://blog.doctoroz.com/dr-oz-blog/agave-why-we-were-wrong. Accessed 2015 Apr 9.

5. Gardner C et al. Nonnutritive sweeteners: current use and health perspectives: a scientific statement from the American Heart Association and the American Diabetes Association. Circulation. 2012 Jul 24; 126(4):509–19.

6. American Diabetes Association [Internet]. Low calorie sweeteners. Edited 2014 Dec 16. Available from: http://www.diabetes.org/food-and-fitness/food/what-can-i-eat/understanding-carbohydrates/artificial-sweeteners. Accessed 2015 Apr 12.

7. Stellman SD, Garfinkel L. Artificial sweetener use and one-year weight change among women. Prev Med. 1986 Mar; 15(2);195–202.

8. Fowler SP et al. Fueling the obesity epidemic? Artificially sweetened beverage use and long-term weight gain. Obesity. 2008 Aug; 16(8):1894–900.

9. Gardener H et al. Diet soft drink consumption is associated with an increased risk of vascular events in the Northern Manhattan Study. J Gen Intern Med. 2012 Sep; 27(9):1120–6.

10. Lutsey PL, Steffen LM, Stevens J. Dietary intake and the development of the metabolic syndrome: the Atherosclerosis Risk in Communities Study. Circulation. 2008 Feb 12; 117(6):754–61.

11. Dhingra R, Sullivan L, Jacques PF, Wang TJ, Fox CS, Meigs JB, D'Agostino RB, Gaziano JM, Vasan RS. Soft drink consumption and risk of developing cardiometabolic risk factors and the metabolic syndrome in middle-aged adults in the community. Circulation. 2007 Jul 31; 116(5):480–8.

12. American College of Cardiology. Too many diet drinks may spell heart trouble for older women, study suggests. ScienceDaily [Internet]. 29 March 2014. Available from: http://www.sciencedaily.com/releases/2014/03/140329175110.htm. Accessed 2015 Apr 9.

13. Pepino MY et al. Sucralose affects glycemic and hormonal responses to an oral glucose load. Diabetes Care. 2013 Sep; 36(9):2530–5.

14. Anton SD et al. Effects of stevia, aspartame, and sucrose on food intake, satiety, and postprandial glucose and insulin levels. Appetite. 2010 Aug; 55(1):37–43.

15. Yang Q. Gain weight by 'going diet?' Artificial sweeteners and the neurobiology of sugar cravings. Yale J Biol Med. 2010 Jun; 83(2):101–8.
16. Smeets, PA et al. Functional magnetic resonance imaging of human hypothalamic responses to sweet taste ad calories. Am J Clin Nutr. 2005 Nov; 82(5):1011–6.
17. Bellisle F, Drewnowski A. Intense sweeteners, energy intake and the control of body weight. Eur J Clin Nutr. 2007 Jun; 61(6):691–700.
18. Ebbeling CB et al. A randomized trial of sugar-sweetened beverages and adolescent body weight. N Engl J Med. 2012 Oct 11; 367(15):1407–16.
19. Blackburn GL et al. The effect of aspartame as part of a multidisciplinary weight-control program on short- and long-term control of body weight. Am J Clin Nutr. 1997 Feb; 65(2):409–18.
20. De Ruyter JC et al. A trial of sugar-free or sugar sweetened beverages and body weight in children. NEJM. 2012 Oct 11; 367(15):1397–406.
21. Bes-Rastrollo M et al. Financial conflicts of interest and reporting bias regarding the association between sugar-sweetened beverages and weight gain: a systematic review of systematic reviews. PLoS Med. Dec 2013; 10(12) e1001578 doi: 10.1371/journal.pmed.1001578. Accessed 2015 Apr 8.

Chapter 16: Carbohydrates and Protective Fiber

1. Data source for Figure 16.1: Cordain L, Eades MR, Eades MD. Hyperinsulinemic diseases of civilization: more than just Syndrome X. Comparative Biochemistry and Physiology: Part A. 2003; 136:95–112. Available from: http://www.direct-ms.org/sites/default/files/Hyperinsulinemia.pdf. Accessed 2015 Apr 15.
2. Fan MS et al. Evidence of decreasing mineral density in wheat grain over the last 160 years. J Trace Elem Med Biol. 2008; 22(4):315–24. Doi: 10.1016/j.jtemb.2008.07.002. Accessed 2015 Apr 8.
3. Rubio-Tapia A et al. Increased prevalence and mortality in undiagnosed celiac disease. Gastroenterology. 2009 Jul; 137(1):88–93.
4. Thornburn A, Muir J, Proietto J. Carbohydrate fermentation decreases hepatic glucose output in healthy subjects. Metabolism. 1993 Jun; 42(6):780–5.
5. Trout DL, Behall KM, Osilesi O. Prediction of glycemic index for starchy foods. Am J Clin Nutr. 1993 Dec; 58(6):873–8.
6. Jeraci JL. Interaction between human gut bacteria and fibrous substrates. In: Spiller GA, ed. CRC handbook of dietary fiber in human nutrition. Boca Raton, FL: CRC Press, 1993. p. 648.

7. Wisker E, Maltz A, Feldheim W. Metabolizable energy of diets low or high in dietary fiber from cereals when eaten by humans. J Nutr. 1988 Aug; 118(8):945–52.

8. Eaton SB, Eaton SB 3rd, Konner MJ, Shostak M. An evolutionary perspective enhances understanding of human nutritional requirements. J Nutr. 1996 Jun; 126(6): 1732–40.

9. Trowell H. Obesity in the Western world. Plant foods for man. 1975; 1:157–68.

10. U.S. Department of Agriculture ARS. CSFII/DHKS data set and documentation: the 1994 Continuing Survey of Food Intakes by Individuals and the 1994–96 Diet and Health Knowledge Survey. Springfield, VA: National Technical Information Service; 1998.

11. Krauss RM et al. Dietary guidelines for healthy American adults. Circulation. 1996 Oct 1; 94(7):1795–1899.

12. Fuchs CS et al. Dietary fiber and the risk of colorectal cancer and adenoma in women. N Engl J Med. 1999 Jan 21; 340(3):169–76.

13. Alberts DS et al. Lack of effect of a high-fiber cereal supplement on the recurrence of colorectal adenomas. N Engl J Med; 2000 Apr 20; 342(16):1156–62.

14. Burr ML et al. Effects of changes in fat, fish and fibre intakes on death and myocardial reinfarction: diet and reinfarction trial (DART). Lancet. 1989 Sep 30; 2(8666):757–61.

15. Estruch R. Primary prevention of cardiovascular disease with a Mediterranean diet. N Engl J Med. 2013 Apr 4; 368(14):1279-90.

16. Miller WC et al. Dietary fat, sugar, and fiber predict body fat content. J Am Diet Assoc. 1994 Jun; 94(6):612–5.

17. Nelson LH, Tucker LA. Diet composition related to body fat in a multivariate study of 203 men. J Am Diet Assoc. 1996 Aug; 96(8):771–7.

18. Gittelsohn J et al. Specific patterns of food consumption and preparation are associated with diabetes and obesity in a native Canadian community. J Nutr. 1998 Mar; 128(3):541–7.

19. Ludwig DS et al. Dietary fiber, weight gain, and cardiovascular disease risk factors in young adults. JAMA. 1999 Oct 27; 282(16):1539–46.

20. Pereira MA, Ludwig DS. Dietary fiber and body-weight regulation. Pediatric Clin North America. 2001 Aug; 48(4):969–80.

21. Chandalia M et al. Beneficial effects of high fibre intake in patients with type 2 diabetes mellitus. NEJM. 2000 May 11; 342(19):1392–8.

22. Liese AD et al. Dietary glycemic index and glycemic load, carbohydrate and fiber intake, and measure of insulin sensitivity, secretion and adiposity in the Insulin Resistance Atherosclerosis Study. Diab. Care. 2005 Dec; 28(12):2832–8.
23. Schulze MB et al. Glycemic index, glycemic load, and dietary fiber intake and incidence of type 2 diabetes in younger and middle-aged women. Am J Clin Nutr. 2004 Aug; 80(2):348–56.
24. Salmerón J et al. JAMA. Dietary fiber, glycemic load, and risk of non-insulin-dependent diabetes mellitus in women. 1997 Feb 12; 277(6):472–7.
25. Salmerón J et al. Dietary fiber, glycemic load, and risk of NIDDM in men. Diabetes Care. 1997 Apr; 20(4):545–50.
26. Kolata G. Rethinking thin: the new science of weight loss—and the myths and realities of dieting. New York: Picador; 2007.
27. Johnston CS, Kim CM, Buller AJ. Vinegar improves insulin sensitivity to a high-carbohydrate meal in subjects with insulin resistance or type 2 diabetes. Diabetes Care. 2004 Jan; 27(1):281–2.
28. Johnston CS et al. Examination of the antiglycemic properties of vinegar in healthy adults. Ann Nutr Metab. 2010; 56(1):74–9. doi 10.1159/0002722133. Accessed 2015 Apr 8.
29. Sugiyama M et al. Glycemic index of single and mixed meal foods among common Japanese foods with white rice as a reference food. European Journal of Clinical Nutrition. 2003 Jun; 57(6):743–752.
30. Ostman EM et al. Inconsistency between glycemic and insulinemic responses to regular and fermented milk products. Am J Clin Nutr. 2001 Jul; 74(1):96–100.
31. Leeman M et al. Vinegar dressing and cold storage of potatoes lowers post-prandial glycaemic and insulinaemic responses in healthy subjects. Eur J Clin Nutr. 2005 Nov; 59(11):1266–71.
32. White AM, Johnston CS. Vinegar ingestion at bedtime moderates waking glucose concentrations in adults with well-controlled type 2 diabetes. Diabetes Care. 2007 Nov; 30(11):2814–5.
33. Johnston CS, Buller AJ. Vinegar and peanut products as complementary foods to reduce postprandial glycemia. J Am Diet Assoc. 2005 Dec; 105(12):1939–42.
34. Brighenti F et al. Effect of neutralized and native vinegar on blood glucose and acetate responses to a mixed meal in healthy subjects. Eur J Clin Nutr. 1995 Apr; 49(4):242–7.

35. Hu FB et al. Dietary intake of a-linolenic acid and risk of fatal ischemic heart disease among women. Am J Clin Nutr. 1999 May; 69(5):890–7.

Chapter 17: Protein

1. Friedman et al. Comparative effects of low-carbohydrate high-protein versus low-fat diets on the kidney. Clin J Am Soc Nephrol. 2012 Jul; 7(7):1103–11.
2. Holt SH et al. An insulin index of foods: the insulin demand generated by 1000-kJ portions of common foods. Am J Clin Nutr. 1997 Nov; 66(5):1264–76.
3. Floyd JC Jr. Insulin secretion in response to protein ingestion. J Clin Invest. 1966 Sep; 45(9):1479-1486
4. Nuttall FQ, Gannon MC. Plasma glucose and insulin response to macronutrients in non diabetic and NIDDM subjects. Diabetes Care. 1991 Sep; 14(9):824–38.
5. Nauck M et al. Reduced incretin effect in type 2 (non-insulin-dependent) diabetes. Diabetologia. 1986 Jan; 29(1):46–52.
6. Pepino MY et al. Sucralose affects glycemic and hormonal responses to an oral glucose load. Diabetes Care. 2013 Sep; 36(9):2530–5.
7. Just T et al. Cephalic phase insulin release in healthy humans after taste stimulation? Appetite. 2008 Nov; 51(3):622–7.
8. Nilsson M et al. Glycemia and insulinemia in healthy subjects after lactose equivalent meals of milk and other food proteins. Am J Clin Nutr. 2004 Nov; 80(5):1246–53.
9. Liljeberg EH, Bjorck I. Milk as a supplement to mixed meals may elevate postprandial insulinaemia. Eur J Clin Nutr. 2001 Nov; 55(11):994–9.
10. Nilsson M et al. Glycemia and insulinemia in healthy subjects after lactose-equivalent meals of milk and other food proteins: the role of plasma amino acids and incretins. Am J Clin Nutr. 2004 Nov; 80(5):1246–53.
11. Jakubowicz D, Froy O, Ahrén B, Boaz M, Landau Z, Bar-Dayan Y, Ganz T, Barnea M, Wainstein J. Incretin, insulinotropic and glucose-lowering effects of whey protein pre-load in type 2 diabetes: a randomized clinical trial. Diabetologia. Sept 2014; 57(9):1807–11.
12. Pal S, Ellis V. The acute effects of four protein meals on insulin, glucose, appetite and energy intake in lean men. Br J Nutr. 2010 Oct; 104(8):1241–48.
13. Data source for Figure 17.1: Ibid.
14. Bes-Rastrollo M, Sanchez-Villegas A, Gomez-Gracia E, Martinez JA, Pajares RM, Martinez-Gonzalez MA. Predictors of weight gain in a Mediterranean cohort:

the Seguimiento Universidad de Navarra Study 1. Am J Clin Nutr. 2006 Feb; 83(2):362–70.

15. Vergnaud AC et al. Meat consumption and prospective weight change in participants of the EPIC-PANACEA study. Am J Clin Nutr. 2010 Aug; 92(2):398–407.
16. Rosell M et al. Weight gain over 5 years in 21,966 meat-eating, fish-eating, vegetarian, and vegan men and women in EPIC-Oxford. Int J Obes (Lond). 2006 Sep; 30(9):1389–96.
17. Mozaffarian D et al. Changes in diet and lifestyle and long-term weight gain in women and men. N Engl J Med. 2011 Jun 23; 364(25):2392–404.
18. Cordain L et al. Fatty acid analysis of wild ruminant tissues: evolutionary implications for reducing diet-related chronic disease. Eur J Clin Nutr. 2002 Mar; 56(3):181–91.
19. Rosell M et al. Association between dairy food consumption and weight change over 9 y in 19,352 perimenopausal women. Am J Clin Nutr. 2006 Dec; 84(6):1481–8.
20. Pereira MA et al. Dairy consumption, obesity, and the insulin resistance syndrome in young adults: the CARDIA Study. JAMA. 2002 Apr 24; 287(16):2081–9.
21. Choi HK et al. Dairy consumption and risk of type 2 diabetes mellitus in men: a prospective study. Arch Intern Med. 2005 May 9; 165(9):997–1003.
22. Azadbakht L et al. Dairy consumption is inversely associated with the prevalence of the metabolic syndrome in Tehranian adults. Am J Clin Nutr. 2005 Sep; 82(3):523–30.
23. Mozaffarian D et al. Changes in diet and lifestyle and long-term weight gain in women and men. N Engl J Med. 2011 Jun 23; 364(25):2392–404.
24. Burke LE et al. A randomized clinical trial testing treatment preference and two dietary options in behavioral weight management: preliminary results of the impact of diet at 6 months—PREFER study. Obesity (Silver Spring). 2006 Nov; 14(11):2007–17.

Chapter 18: Fat Phobia

1. Keys A. Mediterranean diet and public health: personal reflections. Am J Clin Nutr. 1995 Jun; 61(6 Suppl):1321S–3S.
2. Nestle M. Mediterranean diets: historical and research overview. Am J Clin Nutr. 1995 June; 61(6 suppl):1313S –20S.
3. Keys A, Keys M. Eat well and stay well. New York: Doubleday & Company; 1959. p. 40.

4. U.S. Department of Agriculture, U.S. Department of Health and Human Services. Nutrition and your health: dietary guidelines for Americans. 3rd ed. Washington, DC: US Government Printing Office; 1990.

5. The Seven Countries Study. Available from www.sevencountriesstudy.com. Accessed 2015 Apr 12.

6. Howard BV et al. Low fat dietary pattern and risk of cardiovascular disease: the Womens' Health Initiative Randomized Controlled Dietary Modification Trial. JAMA. 2006 Feb 8; 295(6):655–66.

7. Yerushalmy J, Hilleboe HE. Fat in the diet and mortality from heart disease: a methodologic note. N Y State J Med. 1957 Jul 15; 57(14):2343–54.

8. Pollan, Michael. Unhappy meals. New York Times [Internet]. 2007 Jan 28. Available from: http://www.nytimes.com/2007/01/28/magazine/28nutritionism.t.html?-pagewanted=all. Accessed 2015 Sep 6.

9. Simopoulos AP. Omega-3 fatty acids in health and disease and in growth and development. Am J Clin Nutr. 1991 Sep; 54(3):438–63.

10. Eades M. Framingham follies. The Blog of Michael R. Eades, M.D. [Internet]. 2006 Sep 28. Available from: http://www.proteinpower.com/drmike/cardiovascu-lar-disease/framingham-follies/. Accessed 2015 Apr 12.

11. Nichols AB et al. Daily nutritional intake and serum lipid levels. The Tecumseh study. Am J Clin Nutr. 1976 Dec; 29(12):1384–92.

12. Garcia-Pamieri et al. Relationship of dietary intake to subsequent coronary heart disease incidence: The Puerto Rico Heart Health Program. Am J Clin Nutr. 1980 Aug; 33(8):1818–27.

13. Shekelle RB et al. Diet, serum cholesterol, and death from coronary disease: the Western Electric Study. N Engl J Med. 1981 Jan 8; 304(2):65–70.

14. Aro A et al. Transfatty acids in dairy and meat products from 14 European countries: the TRANSFAIR Study. Journal of Food Composition and Analysis. 1998 Jun; 11(2):150–160. doi: 10.1006/jfca.1998.0570. Accessed 2015 Apr 12.

15. Mensink RP, Katan MB. Effect of dietary trans fatty acids on high-density and low-density lipoprotein cholesterol levels in healthy subjects. N Engl J Med. 1990 Aug 16; 323(7):439–45.

16. Mozaffarian D et al. Trans fatty acids and cardiovascular disease. N Engl J Med. 2006 Apr 13; 354(15):1601–13.

17. Mente A et al. A systematic review of the evidence supporting a causal link between dietary factors and coronary heart disease. Arch Intern Med. 2009 Apr 13; 169(7):659–69.

18. Hu FB et al. Dietary fat intake and the risk of coronary heart disease in women. N Engl J Med. 1997 Nov 20; 337(21):1491–9.
19. Leosdottir M et al. Dietary fat intake and early mortality patterns: data from the Malmo Diet and Cancer Study. J Intern Med. 2005 Aug; 258(2):153–65.
20. Chowdhury R et al. Association of dietary, circulating, and supplement fatty acids with coronary risk: a systematic review and meta-analysis. Ann Intern Med. 2014 Mar 18; 160(6):398–406.
21. Siri-Tarino PW et al. Meta-analysis of prospective cohort studies evaluating the association of saturated fat with cardiovascular disease. Am J Clin Nutr. 2010 Mar; 91(3):535–46.
22. Yamagishi K et al. Dietary intake of saturated fatty acids and mortality from cardiovascular disease in Japanese. Am J Clin Nutr. First published 2010 August 4. doi: 10.3945/ ajcn.2009.29146. Accessed 2015 Apr 12.
23. Wakai K et al. Dietary intakes of fat and total mortality among Japanese populations with a low fat intake: the Japan Collaborative Cohort (JACC) Study. Nutr Metab (Lond). 2014 Mar 6; 11(1):12.
24. Ascherio A et al. Dietary fat and risk of coronary heart disease in men: cohort follow up study in the United States. BMJ. 1996 Jul 13; 313(7049):84–90.
25. Gillman MW et al. Margarine intake and subsequent heart disease in men. Epidemiology. 1997 Mar; 8(2):144–9.
26. Mozaffarian D et al. Dietary fats, carbohydrate, and progression of coronary atherosclerosis in postmenopausal women. Am J Clin Nutr. 2004 Nov; 80(5):1175–84.
27. Kagan A et al. Dietary and other risk factors for stroke in Hawaiian Japanese men. Stroke. 1985 May–Jun; 16(3):390–6.
28. Gillman MW et al. Inverse association of dietary fat with development of ischemic stroke in men. JAMA. 1997 Dec 24–31; 278(24):2145–50.
29. National Cholesterol Education Program Expert Panel on Detection, Evaluation, and Treatment of High Blood Cholesterol in Adults (Adult Treatment Panel III). National Institutes of Health; National Heart, Lung, and Blood Institute. 2002 Sep. Available from: http://www.nhlbi.nih.gov/files/docs/resources/heart/atp3full.pdf. Accessed 2015 Apr 12.
30. Kratz M et al. The relationship between high-fat dairy consumption and obesity, cardiovascular, and metabolic disease. Eur J Nutr. 2013 Feb; 52(1):1–24.
31. Rosell M et al. Association between dairy food consumption and weight change over 9 y in 19,352 perimenopausal women. Am J Clin Nutr. 2006 Dec; 84(6):1481–8.

32. Collier G, O'Dea K. The effect of co-ingestion of fat on the glucose, insulin and gastric inhibitory polypeptide responses to carbohydrate and protein. Am J Clin Nutr. 1983 Jun; 37(6):941–4.

33. Willett WC. Dietary fat plays a major role in obesity: no. Obes Rev. 2002 May; 3(2):59–68.

34. Howard BV et al. Low fat dietary pattern and risk of cardiovascular disease. JAMA. 2006 Feb 8; 295(6):655–66.

Chapter 19: What to Eat

1. Knowler WC et al. 10-year follow-up of diabetes incidence and weight loss in the Diabetes Prevention Program Outcomes Study. Lancet. 2009 Nov 14; 374(9702):1677–86.

2. Leibel RL, Hirsch J. Diminished energy requirements in reduced-obese patients. Metabolism. 1984 Feb; 33(2):164–70.

3. Sacks FM et al. Comparison of weight-loss diets with different compositions of fat, protein, and carbohydrates. N Engl J Med. 2009 Feb 26; 360(9):859–73.

4. Johnston BC et al. Comparison of weight loss among named diet programs in overweight and obese adults: a meta-analysis. JAMA. 2014 Sep 3; 312(9):923–33.

5. Grassi D, Necozione S, Lippi C, Croce G, Valeri L, Pasqualetti P, Desideri G, Blumberg JB, Ferri C. Cocoa reduces blood pressure and insulin resistance and improves endothelium-dependent vasodilation in hypertensives. Hypertension. 2005 Aug; 46(2):398–405.

6. Grassi D et al. Blood pressure is reduced and insulin sensitivity increased in glucose-intolerant, hypertensive subjects after 15 days of consuming high-polyphenol dark chocolate. J. Nutr. 2008 Sep; 138(9):1671–6.

7. Djousse L et al. Chocolate consumption is inversely associated with prevalent coronary heart disease: the National Heart, Lung, and Blood Institute Family Heart Study. Clin Nutr. 2011 Apr; 30(2):182–7. doi: 10.1016/j.clnu.2010.08.005. Epub 2010 Sep 19. Accessed 2015 Apr 6.

8. Sabate J, Wien M. Nuts, blood lipids and cardiovascular disease. Asia Pac J Clin Nutr. 2010; 19(1):131–6.

9. Jenkins DJ et al. Possible benefit of nuts in type 2 diabetes. J. Nutr. 2008 Sep; 138(9):1752S–1756S.

10. Hernandez-Alonso P et al. Beneficial effect of pistachio consumption on glucose metabolism, insulin resistance, inflammation, and related metabolic risk markers:

a randomized clinical trial. 2014 Aug 14. doi: 10.2337/dc14-1431. [Epub ahead of print] Accessed 2015 Apr 6.

11. Walton AG. All sugared up: the best and worst breakfast cereals for kids. Forbes [Internet]. 2014 May 15. Available at: http://www.forbes.com/sites/alicegwalton/2014/05/15/all-sugared-up-the-best-and-worst-breakfast-cereals-for-kids/. Accessed 2015 Apr 12.
12. Fernandez ML. Dietary cholesterol provided by eggs and plasma lipoproteins in healthy populations. Curr Opin Clin Nutr Metab Care. 2006 Jan; 9(1):8–12.
13. Mutungi G et al. Eggs distinctly modulate plasma carotenoid and lipoprotein subclasses in adult men following a carbohydrate-restricted diet. J Nutr Biochem. 2010 Apr; 21(4):261–7. doi: 10.1016/j.jnutbio.2008.12.011. Epub 2009 Apr 14.
14. Shin JY, Xun P, Nakamura Y, He K. Egg consumption in relation to risk of cardiovascular disease and diabetes: a systematic review and meta-analysis. Am J Clin Nutr. 2013 Jul; 98(1):146–59.
15. Rong Y et al. Egg consumption and risk of coronary heart disease and stroke: dose-response meta-analysis of prospective cohort studies. BMJ. 2013; 346:e8539. doi: 10.1136/bmj.e8539. Accessed 2015 Apr 6.
16. Cordain L et al. Influence of moderate chronic wine consumption on insulin sensitivity and other correlates of syndrome X in moderately obese women. Metabolism. 2000 Nov; 49(11):1473–8.
17. Cordain L et al. Influence of moderate daily wine consumption on body weight regulation and metabolism in healthy free-living males. J Am Coll Nutr. 1997 Apr; 16(2):134–9.
18. Napoli R et al. Red wine consumption improves insulin resistance but not endothelial function in type 2 diabetic patients. Metabolism. 2005 Mar; 54(3):306–13.
19. Huxley R et al. Coffee, decaffeinated coffee, and tea consumption in relation to incident type 2 diabetes mellitus: a systematic review with meta-analysis. Arch Intern Med. 2009 Dec 14; 169(22):2053–63.
20. Gómez-Ruiz JA, Leake DS, Ames JM. In vitro antioxidant activity of coffee compounds and their metabolites. J Agric Food Chem. 2007 Aug 22; 55(17):6962–9.
21. Milder IE, Arts I, Cvan de Putte B, Venema DP, Hollman PC. Lignan contents of Dutch plant foods: a database including lariciresinol, pinoresinol, secoisolariciresinol and metairesinol. Br J Nutr. 2005 Mar; 93(3):393–402.
22. Clifford MN. Chlorogenic acids and other cinnamates: nature, occurrence and dietary burden. J Sci Food Agric. 1999; 79(5):362–72.

23. Huxley R et al. Coffee, decaffeinated coffee, and tea consumption in relation to incident type 2 diabetes mellitus: a systematic review with meta-analysis. Arch Intern Med. 2009 Dec 14; 169(22):2053–63.

24. Van Dieren S et al. Coffee and tea consumption and risk of type 2 diabetes. Diabetologia. 2009 Dec; 52(12):2561–9.

25. Odegaard AO et al. Coffee, tea, and incident type 2 diabetes: the Singapore Chinese Health Study. Am J Clin Nutr. 2008 Oct; 88(4):979–85.

26. Freedman ND, Park Y, Abnet CC, Hollenbeck AR, Sinha R. Association of coffee drinking with total and cause-specific mortality. N Engl J Med. 2012 May 17; 366(20):1891–904.

27. Lopez-Garcia E, van Dam RM, Li TY, Rodriguez-Artalejo F, Hu FB. The relationship of coffee consumption with mortality. Ann Intern Med. 2008 Jun 17; 148(2):904–14.

28. Eskelinen MH, Kivipelto M. Caffeine as a protective factor in dementia and Alzheimer's disease. J Alzheimers Dis. 2010; 20 Suppl 1:167–74.

29. Santos C et al. Caffeine intake and dementia: systematic review and meta-analysis. J Alzheimers Dis. 2010; 20 Suppl 1:S187–204. doi: 10.3233/JAD-2010-091387. Accessed 2015 Apr 6.

30. Hernan MA et al. A meta-analysis of coffee drinking, cigarette smoking, and the risk of Parkinson's disease. Ann Neurol. 2002 Sep; 52(3):276–84.

31. Ross GW et al. Association of coffee and caffeine intake with the risk of Parkinson disease. JAMA. 2000 May; 283(20):2674–9.

32. Klatsky AL et al. Coffee, cirrhosis, and transaminase enzymes. Arch Intern Med. 2006 Jun 12; 166(11):1190–5.

33. Larrson SC, Wolk A. Coffee consumption and risk of liver cancer: a meta-analysis. Gastroenterology. 2007 May; 132 (5):1740–5.

34. Kobayashi Y, Suzuki M, Satsu H et al. Green tea polyphenols inhibit the sodium-dependent glucose transporter of intestinal epithelial cells by a competitive mechanism. J Agric Food Chem. 2000 Nov; 48(11):5618–23.

35. Crespy V, Williamson GA. A review of the health effects of green tea catechins in in vivo animal models. J Nutr. 2004 Dec; 134(12 suppl):3431S–3440S.

36. Cabrera C et al. Beneficial effects of green tea: a review. J Am Coll Nutr. 2006 Apr; 25(2):79–99.

37. Hursel, R, Westerterp-Plantenga MS. Catechin- and caffeine-rich teas for control of body weight in humans. Am J Clin Nutr. 2013 Dec; 98(6):1682S–93S.

38. Dulloo AG et al. Green tea and thermogenesis: interactions between catechin-polyphenols, caffeine and sympathetic activity. Inter J Obesity. 2000 Feb; 24(2):252–8.

39. Venables MC et al. Green tea extract ingestion, fat oxidation, and glucose tolerance in healthy humans. Am J Clin Nutr. 2008 Mar; 87(3):778–84.

40. Dulloo AG et al. Efficacy of a green tea extract rich in catechin polyphenols and caffeine in increasing 24-h energy expenditure and fat oxidation in humans. Am J Clin Nutr. 1999 Dec; 70(6):1040–5.

41. Koo MWL, Cho CH. Pharmacological effects of green tea on the gastrointestinal system. Eur J Pharmacol. 2004 Oct 1; 500(1-3):177–85.

42. Hursel R Viechtbauer W, Westerterp-Plantenga, MS. The effects of green tea on weight loss and weight maintenance: a meta-analysis. Int J Obes (Lond). 2009 Sep; 33(9):956–61. doi: 10.1038/ijo.2009.135. Epub 2009 Jul 14. Accessed 6 Apr 2015.

43. Van Dieren S et al. Coffee and tea consumption and risk of type 2 diabetes. Diabetologia. 2009 Dec; 52(12):2561–9.

44. Odegaard, AO et al. Coffee, tea, and incident type 2 diabetes: the Singapore Chinese Health Study. Am J Clin Nutr. 2008 Oct; 88(4):979–85.

45. Patrick L, Uzick M. Cardiovascular disease: C-reactive protein and the inflammatory disease paradigm: HMG-CoA reductase inhibitors, alpha-tocopherol, red yeast rice, and olive oil polyphenols. A review of the literature. Alternative Medicine Review. 2001 Jun; 6(3):248–71.

46. Aviram M, Eias K. Dietary olive oil reduces low-density lipoprotein uptake by macrophages and decreases the susceptibility of the lipoprotein to undergo lipid peroxidation. Ann Nutr Metab. 1993; 37(2):75–84.

47. Smith RD et al. Long-term monounsaturated fatty acid diets reduce platelet aggregation in healthy young subjects. Br J Nutr. 2003 Sep; 90(3):597–606.

48. Ferrara LA et al. Olive oil and reduced need for antihypertensive medications. Arch Intern Med. 2000 Mar 27; 160(6):837–42.

49. Martínez-González MA et al. Olive oil consumption and risk of CHD and/or stroke: a meta-analysis of case-control, cohort and intervention studies. Br J Ntru. 2014 Jul; 112(2):248–59.

50. Chen M, Pan A, Malik VS, Hu FB. Effects of dairy intake on body weight and fat: a meta-analysis of randomized controlled trials. Am J Clin Nutr. 2012 Oct; 96(4):735–47.

51. Mozaffarian, D et al. Trans-palmitoleic acid, metabolic risk factors, and new-onset diabetes in U.S. adults: a cohort study. Ann Intern Med. 2010 Dec 21; 153(12):790–9.

52. Hyman M. The super fiber that controls your appetite and blood sugar. Huffington Post [Internet]. 2010 May 29 (updated 2013 Nov 11). Available from: http://www.huffingtonpost.com/dr-mark-hyman/fiber-health-the-super-fi_b_594153.html. Accessed 2015 Apr 6.

53. Sugiyama M et al. Glycemic index of single and mixed meal foods among common Japanese foods with white rice as a reference food. Euro J Clin Nutr. 2003 Jun; 57(6):743–52. doi:10.1038/sj.ejcn.1601606. Accessed 2015 Apr 6.

Chapter 20: When to Eat

1. Arbesmann R. Fasting and prophecy in pagan and Christian antiquity. Traditio. 1951; 7:1–71.
2. Lamine F et al. Food intake and high density lipoprotein cholesterol levels changes during Ramadan fasting in healthy young subjects. Tunis Med. 2006 Oct; 84(10):647–650.
3. Felig P. Starvation. In: DeGroot LJ, Cahill GF Jr et al., editors. Endocrinology: Vol 3. New York: Grune & Stratton; 1979. pp. 1927–40.
4. Coffee CJ, Quick look: metabolism. Hayes Barton Press; 2004. p. 169.
5. Owen OE, Felig P. Liver and kidney metabolism during prolonged starvation. J Clin Invest. 1969 Mar; 48:574–83.
6. Merrimee TJ, Tyson JE. Stabilization of plasma glucose during fasting: normal variation in two separate studies. N Engl J Med. 1974 Dec 12; 291(24):1275–8.
7. Heilbronn LK. Alternate-day fasting in nonobese subjects: effects on body weight, body composition, and energy metabolism. Am J Clin Nutr. 2005; 81:69–73.
8. Halberg N. Effect of intermittent fasting and refeeding on insulin action in healthy men. J Appl Physiol. 1985 Dec; 99(6):2128–36.
9. Rudman D et al. Effects of human growth hormone in men over 60 years old. N Engl J Med. 1990 Jul 5; 323(1):1–6.
10. Ho KY et al. Fasting enhances growth hormone secretion and amplifies the complex rhythms of growth hormone secretion in man. J Clin Invest. 1988 Apr; 81(4):968–75.
11. Drenick EJ. The effects of acute and prolonged fasting and refeeding on water, electrolyte, and acid-base metabolism. In: Maxwell MH, Kleeman CR, editors. Clinical disorders of fluid and electrolyte metabolism. 3rd ed. New York: McGraw-Hill; 1979.

12. Kerndt PR et al. Fasting: the history, pathophysiology and complications. West J Med. 1982 Nov; 137(5):379–99.
13. Stewart WK, Fleming LW. Features of a successful therapeutic fast of 382 days' duration. Postgrad Med J. 1973 Mar; 49(569):203–9.
14. Lennox WG. Increase of uric acid in the blood during prolonged starvation. JAMA. 1924 Feb 23; 82(8):602–4.
15. Drenick EJ et al. Prolonged starvation as treatment for severe obesity. JAMA. 1964 Jan 11; 187:100–5.
16. Felig P. Starvation. In: DeGroot LJ, Cahill GF Jr et al., editors. Endocrinology: Vol 3. New York: Grune & Stratton; 1979. pp. 1927–40.
17. Bhutani S et al. Improvements in coronary heart disease risk indicators by alternate-day fasting involve adipose tissue modulations. Obesity. 2010 Nov; 18(11):2152–9.
18. Stote KS et al. A controlled trial of reduced meal frequency without caloric restriction in healthy, normal-weight, middle-aged adults. Am J Clin Nutr. 2007 Apr; 85(4):981–8.
19. Heilbronn LK. Alternate-day fasting in nonobese subjects: effects on body weight, body composition, and energy metabolism. Am J Clin Nutr. 2005; 81:69–73.
20. Zauner C. Resting energy expenditure in short-term starvation is increased as a result of an increase in serum norepinephrine. Am J Clin Nutr. 2000 Jun; 71(6):1511–5.
21. Stubbs RJ et al. Effect of an acute fast on energy compensation and feeding behaviour in lean men and women. Int J Obesity. 2002 Dec; 26(12):1623–8.
22. Duncan GG. Intermittent fasts in the correction and control of intractable obesity. Trans Am Clin Climatol Assoc 1963; 74:121–9.
23. Duncan DG et al. Correction and control of intractable obesity. Practical application of Intermittent Periods of Total Fasting. JAMA. 1962; 181(4):309–12.
24. Drenick E. Prolonged starvation as treatment for severe obesity. JAMA. 1964 Jan 11; 187:100–5.
25. Thomson TJ et al. Treatment of obesity by total fasting for up to 249 days. Lancet. 1966 Nov 5; 2(7471):992–6.
26. Kerndt PR et al. Fasting: the history, pathophysiology and complications. West J Med. 1982 Nov; 137(5):379–99.
27. Folin O, Denis W. On starvation and obesity, with special reference to acidosis. J Biol Chem. 1915; 21:183–92.

28. Bloom WL. Fasting as an introduction to the treatment of obesity. Metabolism. 1959 May; 8(3):214–20.

29. Stewart WK, Fleming LW. Features of a successful therapeutic fast of 382 days' duration. Postgrad Med J. 1973 Mar; 49(569):203–9.

30. Merimee TJ, Tyson JE. Stabilization of plasma glucose during fasting: Normal variation in two separate studies. N Engl J Med. 1974 Dec 12; 291(24):1275–8.

31. Bloom WL. Fasting ketosis in obese men and women. J Lab Clin Med. 1962 Apr; 59:605–12.

32. Forbes GB. Weight loss during fasting: implications for the obese. Am J Clin Nutr. 1970 Sep; 23:1212–19.

33. Harvie MN et al. The effects of intermittent or continuous energy restriction on weight loss and metabolic disease risk markers. Int J Obes (Lond). 2011 May; 35(5):714–27.

34. Klempel MC et al. Intermittent fasting combined with calorie restriction is effective for weight loss and cardio-protection in obese women. Nutr J. 2012; 11:98. doi: 10.1186/1475-2891-11-98. Accessed 2015 Apr 8.

35. Williams KV et al. The effect of short periods of caloric restriction on weight loss and glycemic control in type 2 diabetes. Diabetes Care. 1998 Jan; 21(1):2–8.

36. Koopman KE et al. Hypercaloric diets with increased meal frequency, but not meal size, increase intrahepatic triglycerides: A randomized controlled trial. Hepatology. 2014 Aug; 60(2); 545–55.

37. Yanovski JA, Yanovski SZ, Sovik KN, Nguyen TT, O'Neil PM, Sebring NG. A prospective study of holiday weight gain. N Engl J Med. 2000 Mar 23; 342(12):861–7.

Appendix B

1. Hiebowicz J et al. Effect of cinnamon on post prandial blood glucose, gastric emptying and satiety in healthy subjects. Am J Clin Nutr. 2007 Jun; 85(6):1552–6.

2. Greenberg JA, Geliebter A. Coffee, hunger, and peptide YY. J Am Coll Nutr. 2012 Jun; 31(3):160–6.

INDEX

•

Figures indicated by page numbers in italics.